B

ISNM 77:
International Series of Numerical Mathematics
Internationale Schriftenreihe zur Numerischen Mathematik
Série internationale d'Analyse numérique
Vol. 77

Edited by
Ch. Blanc, Lausanne; R. Glowinski, Paris;
G. Golub, Stanford; P. Henrici, Zürich;
H. O. Kreiss, Pasadena; A. Ostrowski, Montagnola;
J. Todd, Pasadena

Birkhäuser Verlag
Basel · Boston · Stuttgart

Inverse Problems

Proceedings of the Conference held at
the Mathematical Research Institute at Oberwolfach,
Black Forest, May 18–24, 1986

Edited by

John Rozier Cannon
Ulrich Hornung

1986

Birkhäuser Verlag
Basel · Boston · Stuttgart

Editors

Prof. Dr. John Rozier Cannon
Dept. Pure & Appl. Mathematics
Washington State University
Pullman, Washington 99163
USA

Prof. Dr. Ulrich Hornung
Sonderforschungsbereich 72
Wegelerstrasse 6
D–5300 Bonn 1

Library of Congress Cataloging in Publication Data

Inverse problems.
    (International series of numerical mathematics ;
vol. 77 = Internationale Schriftenreihe zur numerischen
mathematik = Série internationale d'analyse numérique ;
vol. 77)
    »Lectures given at the Conference on ›Inverse
Problems‹« – – Pref.
    Bibliography: p.
    Includes index.
    1. Inverse problems (Differential equations) – –
Congresses.   I. Cannon, John Rozier, 1938–   .
II. Hornung, Ulrich, 1941–   .   III. Mathematisches
Forschungsinstitut Oberwolfach.   IV. Conference on
»Inverse Problems« (1986 : Mathematical Research
Institute at Oberwolfach)   V. Series: International
series of numerical mathematics ; v. 77.
QA370.I59   1986   515.3'5   86–29904

ISBN-13: 978-3-0348-7016-0   e-ISBN-13: 978-3-0348-7014-6
DOI: 10.1007/978-3-0348-7014-6

CIP-Kurztitelaufnahme der Deutschen Bibliothek

Inverse problems : proceedings of the conference
held at the Math. Research Inst. at Oberwolfach,
Black Forest, May 18–24, 1986 / ed. by John
Rozier Cannon ; Ulrich Hornung. – Basel ; Boston ;
Stuttgart : Birkhäuser, 1986.
    (International series of numerical mathematics ;
    Vol. 77)

NE: Cannon, John Rozier [Hrsg.]; GT

PREFACE

The present volume contains manuscripts of lectures or topics related to
the lectures which were given at the conference on "Inverse Problems" at the
mathematical Research Institute at Oberwolfach. The conference took place
during the week of May 18-24, 1986, and was managed by the editors.

Recalling Professor Joseph Keller's paper entitled Inverse Problems,
American Mathematical Monthly, 83(1976), we give two direct quotes. "We
call two problems _inverses_ of one another if the formulation of each involves
all or part of the solution of the other. Often, for historical reasons,
one of the two problems has been studied extensively for some time, while the
other is newer and not so well understood. In such cases, the former is
called the _direct_ problem, while the latter is called the _inverse_ problem."
"The main sources of inverse problems are science and engineering. Often
these problems concern the determination of the properties of some inaccessible regions from observations on the boundary of that region." Often,
inverse problems are not well posed. This increases the difficulty in their
analysis and numerical solution. As can be seen from the table of content
of this volume, the conference covered inverse problems in scattering theory,
seismology, tomography, estimation of coefficients and source terms in
parabolic and elliptic differential equations, the inverse Sturm-Liouville
problem, and numerical methods.

The editors wish to thank Professor M. Barner and his co-workers of the
Mathematical Research Institute for their help in creating the conference.
The editors also wish to thank Birkhäuser Verlag for their production of this
book which will enable many more scientists to enjoy the stimulation of the
conference.

John Rozier Cannon (Pullman)   Ulrich Hornung (Neubiberg)

C O N T E N T S

List of contributors

R. S. Anderssen
Department of Mathematics
University of Queensland
St. Lucia, Qld 4067
Australia

J. R. Cannon
Dept. Pure & Appl. Math.
Washington State University
Pullman, Washington 99163
U. S. A.

P. C. DuChateau
Department of Mathematics
Colorado State University
Fort Collins, Colorado 80523
U. S. A.

M. Cheney
Department of Mathematics
Duke University
Durham, NC 27706
U. S. A.

K. R. Driessel
AMOCO Production Company
4502 East 41st Street
P. O. Box 591
Tulsa, Oklahoma 74102
U. S. A.

R. Kreß
Institut für Numerische und
Angewandte Mathematik
Georg-August-Universität
Lotzestraße 16-18
D-3400 Göttingen

R. Gorenflo
Fachbereich Mathematik
Freie Universität Berlin
Arnimallee 2-6
D-1000 Berlin 33

R. E. Ewing
The University of Wyoming
Department of Mathematics
Box 3036, University Station
Laramie, WY 82071
U. S. A.

C. D. Pagani
Dipartimento di Matematica
del Politecnico
Piazza Leonardo da Vinci, 32
I-20133 Milano
Italy

S. Pérez-Esteva
Instituto de Matematicas
U. N. A. M.
Ciudad Universitaria
04510 D. F. Mexico

W. Rundell
Department of Mathematics
Texas A & M University
College Station, TX 77843-3368
U. S. A.

T. I. Seidmann
The University of Maryland
Department of Mathematics
Catonsville, Maryland 21228
U. S. A.

T. Suzuki
Department of Mathematics
Faculty of Science
University of Tokyo
Hongo
Tokyo 113
Japan

P. C. Sabatier
Laboratoire de Physique Math.
Univ. des Sciences et Techn.
du Languedoc
Place Bataillon
F-34060 Montpellier-Cedex

International Series of
Numerical Mathematics, Vol. 77
© 1986 Birkhäuser Verlag Basel

THE LINEAR FUNCTIONAL STRATEGY FOR IMPROPERLY POSED PROBLEMS

R.S. Anderssen[1], CSIRO Division of Mathematics and
Statistics, and Centre for Mathematical Analysis,
Australian National University

ABSTRACT

In the construction of approximation procedures for the solution of inverse problems, some form of stabilization is needed if realistic and useful information about the properties of the solutions is to be generated. There are various ways in which such stabilization can be introduced. They include various forms of regularization, contrained optimization, linear programming inversion, and low dimensional parameterization in the characterization of the approximations. An alternative strategy is to simply limit the information determined and used for inference purposes about the solutions to bounded linear functionals defined on them, such as generalized moments.

Even if, when evaluating such a functional, the approximation to the solution is not very stable, the evaluation of a bounded functional will induce its own stabilization of the information being sought. This can be characterized in a number of independent ways. For example, by a statistical analysis of the effect of the functional on randomly perturbed data. A more cogent and useful characterization can be built around the fact that, by exploiting the mathematical structure of the problem under examination, the linear functionals defined on its solutions (the solution-functionals) can be transformed to linear functionals defined on the data (the data-functionals). The major advantage of this approach is that it gives an explicit characterization of the degree of improperly posedness of a solution functional in terms of the operations performed on the data by the data-functional.

The ill-posedness of the inverse problem has been transferred to the transformation which maps the solution-functionals into the

---

[1]New Address: Department of Mathematics, University of Queensland,
St. Lucia, Qld 4067, Australia.

data-functionals; and has thereby been circumvented, when the transformation can be solved exactly, as long as no attempt is made to reconstruct the solution from the computed estimates of these functionals derived from the available observational data. Even when the transformation must be inverted numerically, the situation is superior to that for the original inverse problem since the data is now known analytically, and therefore the full power of regularization can be exploited.

Thus, from the practical point of view of solving inverse problems which arise in applications, the aim of the linear functional strategy is to identify solution-functionals relevant to the problem context and, where possible, to simply work with them as the corresponding data-functionals for inference and decision making purposes.

§1. INTRODUCTION

From a practical point of view, inverse problems arise in the interpretation of indirect measurements where the aim is to derive information about the phenomenon of interest from the available indirect measurements which have been performed on it. For example, the numerical differentiation of concentration data to determine the rate of change of the concentration in situations where this rate cannot be measured directly (Anderssen and Bloomfield (1974), Figure 1); the determination of the angular distribution of leaves in a canopy of trees from hemispherical lens camera photographs taken of the canopy from below (Anderssen, Jackett and Jupp (1985)); reconstruction from projections in medical tomography applications (Gordon et al (1975)); reconstruction from spectral information (McLaughlin (1986)); the determination of the transmissivity of a confined aquifer from piezometric heat data (Bear (1972), Yeh (1986)).

Within the framework of the formulate → solve (computationally) → interpret model of applied mathematics, the solution of a pracatical inverse problem for which the formulation is already available (as distinct from a purely theoretical examination of the basic mathematical properties of inverse problems) involves both the "solve" and "interpret" steps. These theoretical and practical activities complement each other, but the style of mathematics differs. In the practical context, one aims to exploit the mathematical structure of the specific problem under examination when determining the indicators required for decision making purposes. In the theoretical situation, one tends to study classes of problems with respect to standard properties such as existence, uniqueness, regularity, degree of ill-posedness and stability.

For example, in a practical context, it is not a matter of simply solving the given problem formulation

$$(1) \qquad Ku = s \ ,$$

but a matter of estimating specific properties of u given that the only available information about s is the observational data

$$(2) \qquad d_j = s(x_j) + \epsilon_j, \ j = 1,2,\ldots,J \ ,$$

where the ϵ_j denote observational (random) errors.

For the practitioner, these specific properties will correspond to simple and unambigious indicators which can be used for decision making purposes. In order to be widely accepted and utilized, such indicators must have a clearly defined interpretation in terms of the problem context, have simple algorithms for their evaluation, and be proven to be the indicator on which a particular decison should be based.

The idea of using such indicators as the basis for solving improperly posed problems is not new. It has been suggested in a variety of contexts by a number of authors including Golberg (1979), Pack (1982), Sabatier (1984) and Anderssen (1980). It has been explained succinctly by Sabatier (1984) as <u>the need to identify and ask, within the framework of the indirect measurements, well-posed questions about the phenomenon of interest.</u> For example, the use of signatures in exploration geophysics to locate mineralization requuiring further exploration (Dobrin (1960)); the use of magnetic surveys to locate thick layers of sediments (Koch and Tarlowski (1986)); testing the uniformity of the thickness of coal seams using seismic reflection data (Problem 5 in Barton and Gray (1986); Buchanan (1978)).

The consequences of this approach (the LFS $\equiv$ the linear functional strategy) for the solution of practical inverse problems can be explained as follows. The solution process is similar to the <u>modus operandi</u> for the application of mathematics to industrial problems (Anderssen and de Hoog (1982)). Thus, the application of the LFS reduces the problem solving process to answering the specific questions under examination and nothing else. Moreover, the aim is to do this as simply as possible. Identifying how this might be done is often a non-trivial step requiring insight and understanding about the (mathematical) structure of the specific problem under examination. However, it has far reaching consequences for the solution of practical inverse problems since it is instrumental in creation the situation where many such problems are solved without the assistance of mathematicians and even on occasions without the assistance of much mathematics. This point will be pursued in §2.

§2. THE DIRECT USE OF INDIRECT MEASUREMENTS

In many situations, the type of questions under examination can be answered through <u>the direct use of the available indirect measurements</u> (2). For example, in stereological applications (Moran (1972); Jakeman and Anderssen (1975)), area fraction observed on random plane sections taken through a material yields an (unbiased) estimator of the volume fraction of the components in the material. Such use of the indirect measurements is based on the tacit assumption (which must be formalized and tested) that comparisons of the processes under examination are reflected correctly in corresponding comparisons which can be made on the data. For example, the impact strength of steel can be related to the observations of the carbon particles on random plane sections through the steel; although the process via which steel is strengthened relates to the three-dimensional distribution of the carbon particles in the steel. This example will be pursued further in §3.

In fact, the data being analysed in many industrial laboratories corresponds to the direct manipulation of the indirect measurements; often without a clear appreciation of the phenomenon indirectly under consideration. This is particularly true of situations where the data is analysed on a routine basis as if it where the phenomenon of interest. For example, Anderssen and Campbell (1984) have explained how lower dimensional (stereological) observations of the different components which constitute some biological system such as the cell cycle of Chlorella (Atkinson <u>et al</u>

(1974)) are used to identify and quantify the different stages of the components' contribution to the system's growth and development. Since the same measurement is made on each of the different components, the data themselves yield a direct comparative measure of their growth. Such techniques have been used to show that, among yeasts and Chlorella, the enzymes are usually accumulated discontinuously during the cell division cycle (Atkinson et al (1974)).

An important aspect of the direct use of indirect measurements, which removes deliberations from the normal domain of improperly posed problems, is the fact that, in many industrial and scientific situations, the experimental development of rules of thumb are based on the manipulation of the indirect measurements not the interpretation of the phenomenon under examination. An example is the industrial and scientific use of moiré patterns (Theocaris (1967; 1969)). But even in such situations, there is the need to verify that the measurements performed on the moiré patterns do reflect in some appropriate (approximate) and consistent manner the interpretation being made about the phenomenon of interest.

The importance of this direct use of indirect measurements in the solution of practical inverse problems should not be underestimated. A number of examples have been given above. Another is the method developed by Anderssen and Campbell (1984) for testing for optical symmetry in the (eye) lenses of animals. A more detailed discussion of the topic of this Section and its connection with stereological methods can also be found in their paper.

§3. THE LINEAR FUNCTINAL STRATEGY FOR THE ABEL INTEGRAL EQUATION

As explained in Anderssen and de Hoog (1982), many practical inverse problems reduce to the solution of an Abel integral equation. In paraticular, the Abel equation defines the mathematical framework in which the impact strength of steel example, mentioned in §2, can be analysed.

When random plane sections are taken through a matrix of spherical carbon particles and ferrite crystals, the size (radius) distribution $s(x)$ of the observed circular sections of the carbon particles is related to the size (radius) distribution $u(x)$ of the spherical particles by the Abel integral equation

$$(3) \qquad s(x) = \frac{x}{m} \int_x^a \frac{u(\tau)}{(\tau^2 - x^2)^{1/2}} \, d\tau \ , \ 0 \leq x \leq \tau \leq a \ ,$$

$$(4) \qquad m = \int_0^a x u(x) dx = \pi/2H \ , \ H = \int_0^a \{s(y)/y\} dy \ .$$

Many of the properties of steel, which this matrix of spherical carbon particles and ferrite crystals is modelling, can be derived directly from the observational data

$$d_j = s(x_j) + \epsilon_j \ , \ j = 1,2,\ldots,J \ .$$

Some examples are the comparative assessment of the strengths of different steels; and the determination of the volume fraction of the carbon particles.

However, not all indicators used for inference purposes are necessarily defined on $s(x)$. For example, in many applications, the required indicators correspond to bounded linear functionals defined on $u(\tau)$; viz.

$$(5) \qquad m_\theta(u) = \int_0^a \theta(x) \ u(x) dx \ , \qquad \theta(x) \equiv \text{known} \ .$$

In particular, when testing the impact strength of steel, the indicators commonly used by metallurgists are (Hyam and Nutting (1956)) :

    a. average sphere radius $m = m_x(u)$ as defined in (4) [which must be estimated before (2)-(4) can be solved] ;

    b. average particle surface area

$$(6) \qquad A(u) = 4\pi \int_0^a x^2 u(x) dx \ ;$$

c. average particle volume

$$(7) \qquad V(u) = \frac{4\pi}{3} \int_0^a x^3 u(x)dx \ .$$

One now exploits the fact that various inversion formulas for (3) are known including

$$(8) \qquad u(\tau) = - \frac{2}{\pi} m \frac{d}{d\tau} \left\{ \int_\tau^a (x^2-\tau^2)^{-1/2} s(x)dx \right\} \ .$$

It it is substituted for $u(\tau)$ in (6) and (7) , and the alternative form for m in (4) is used, then the evaluation of the solution functionals m , (6) and (7) is reduced to the evaluation of the following data functionals

$$(9) \qquad m = \frac{\pi}{2} \left\{ \int_0^a \frac{s(x)}{x} dx \right\}^{-1} \ ,$$

$$(10) \qquad A(u) = 16m \int_0^a x \, s(x)dx \ ,$$

$$(11) \qquad V(u) = 2\pi m \int_0^a x^2 \, s(x)dx \ .$$

We see from the form of these data functionals that there is a natural stabilization associated with the application of the linear functional strategy. Whereas the determination of u involves a half-differentiation of $s(x)$, none of the above data functionals involve any differentiation of $s(x)$.

In making this transformation to the corresponding data functionals, a key role has been played by the inversion formula (8) since the differentiation is outside the integral sign. Computationally, this form of the inversion formula tends to be more useful than the others. For example, it is only through this form that Fourier metehods can be used to perform the $\frac{1}{2}$-differentiation when the observations d_j, $j = 1,2,\ldots,J$, are not available on an even grid. A discussion of an incorrect interpretation of an inversion formula for Abel's equation can be found in Anderssen and de Hoog (1981).

A more detailed discussion of the numerical analysis of Abel integral equations can be fouond in Anderssen and de Hoog (1982) and Anderssen (1980) and in the references cited there.

The use of inversion formulas to transform from solutions to data-functionals also arises in the foliage angle distribution problem (Anderssen and Jackett (1984), (1985), Anderssen Jackett and Jupp (1984), and Anderssen et al (1985)).

§4. EXTENSION TO GENERAL FIRST KIND INTEGRAL EQUATIONS

It follows from §3 that a natural way to implement the linear functional strategy is to exploit the mathematical structure of the problem under consideration so that data-functionals corresponding to the required solution-functionals can be identified. The computational efficiency and advantages associated with computing the solution-functionals as their corresponding data-functionals is obvious. The disadvantages are not so obvious and will be discussed below.

In §3. it was shown informally how this identification was more or less automatic for first kind integral equations for which an appropriate inversion formula was available,. Clearly, the integration by parts necessary to accomplish the identification will impose certain regularity conditions on the admissable forms for $\theta(x)$.

However, this identification process is not limited to such first kind equations. It can be generalized to all first kind integral equations in the following manner. Consider the first kind integral operator

$$(12) \qquad Ku = \int_0^1 k(t,\tau)u(\tau)d\tau = s(t) \ , \ K : \underline{D}(K) \to \underline{R}(K) \ ,$$

with

$$(13) \qquad \int_0^1 \int_0^1 |k(t,\tau)|^2 dt d\tau < \infty \ ,$$

and with the domain $\underline{D}(K)$ and range $\underline{R}(K)$ of K contained in a Hilbert space $\underline{H}$ with the L_2-inner product and norm defined by

$$(14) \qquad (u,v) = \int_0^1 uv \ dx, \ \|u\| = (u,u)^{1/2}, \ u,v \in \underline{H} \ .$$

Then the identification of the data-functional corresponding to a given solution-functional can be characterized in the following way (Golberg (1979), Anderssen and de Hoog (1982)).

PROPOSITION. Let K^* denote the adjoint of K with respect to the L_2-inner product (14). If the known θ which defines

$$(15) \qquad m_\theta(u) = \int_0^1 \theta(x)u(x)dx$$

is contained in $\underline{R}(K^*)$, then the required φ which defines

$$(16) \qquad m_\varphi(s) = \int_0^1 \varphi(x)s(x)dx$$

is determined by

$$(17) \qquad K^*\varphi = \theta \ .$$

Immediate consequences of this result are:

(i) It gives an explicit characterization of the degree of improperly posedness of the solution-functional in terms of the operations performed on $s(x)$ by the data-functinal. In most situations, the evaluation of $m_\varphi(s)$ will not involve a differentiation of s , whereas the solution of (12) will often correspond to some (fractional) differentiation of s(cf. Wahba (1980)). This point was illustrated in §3 for the Abel integral equation.

(ii) Formally, the ill-posedness of the inverse problem has been transferred to the transformation (17) which maps the data-functionals into the solution-functionals. Thus, when the transformation (17) can be solved exactly for a given θ , the essential ill-posedness of (12) is circumvented, as long as no attempt is made to reconstruct the solution of (12) from the computed estimates of the solution-functionals derived from the available observational data. However, when the data-functionals cannot be evaluated exactly, the improperly posedness of (12) again plays a role in that some data-functionals determine their corresponding solution-functionals with greater reliability than others. This point will be pursued below.

(iii) Even when the tranformation (17) must be solved numerically for φ , the situation is superior to that for the original inverse problem (12) since the data $\theta(x)$ is now known analyticaly, and therefore the full power of regularization can be exploited. However, there is the usual limitation that the quality with which the structure of φ can be recovered depends on the smoothness of K^*. Thus, the regularized approximation $\hat{\varphi}$ of φ will not contain the high frequency components of φ which the regularization itself has removed in generating the stable approximation $\hat{\varphi}$.

A theoretical consequence of the proposition is the need to identify regularity conditions on $k(t,\tau)$ and $\theta(\tau)$ which guarantee that $\theta(\tau) \in \underline{R}(K^*)$. Some results of this nature can be found in Anderssen (1980). They include: a general characterization for φ when K is a first kind Volterra operator; for first kind Fredholm operators, the identification of conditions on $k(t,\tau)$ for which $\theta(\tau) = \tau^j \notin \underline{R}(K^*)$.

In situations where $\theta(x) \notin \underline{R}(K^*)$, alternative ways for doing the identification can be sought. An example can be found in Anderssen and Jackett (1985), where a general algorithm is developed for computing linear functionals of the foliage angle distribution. It is based on the observation that

$$\int_0^1 \theta(x)u(x)dx = \int_0^1 \theta(x) \frac{d}{dx} \left\{ -\int_x^{\pi/2} u(\tau)d\tau \right\} dx$$

(18)

$$= \left[\theta(x)\int_x^{\pi/2} u(\tau)d\tau\right]_o^1 + \int_o^1 \left[\left\{\int_x^{\pi/2} u(\tau)d\tau\right\} \frac{d\theta(x)}{dx}\right]dx$$

and the fact that a special closed form expression for the indefinite integral $\int_x^{\pi/2} u(\tau)d\tau$ is available.

Even when alternative procedures cannot be found, one can always resort to constructing an approximation $\hat{u}$ to u of (12) and then evaluating $m_\theta(u)$ as $m_\theta(\hat{u})$. For smooth $\theta(x)$, one can appeal to superconvergence arguments or smoothing concepts to infer how the direct evaluation of $m_\theta(u)$ is better posed than the solution of (12).

For the manipulation of bounded linear functionals a variety of techniques is available. They include

The Singular Value Decomposition Interpretation for Linear Functionals

For the $m{\times}n$ matrix A with $m \geq n$, which has the singular value decomposition (Golub and van Loan(1983))

(19)
$$A = U^T \Lambda V$$

where

$$\Lambda = \text{diag } (\sigma_1,\sigma_2,\ldots,\sigma_n), \quad \sigma_j = \text{j-th singular value of } A,$$
$$U \equiv \text{eigenspace of } AA^T,$$
$$V \equiv \text{eigenspace of } A^TA ,$$

it follows that

(20)
$$Au = s$$

can be reinterpreted as the following identification between solution-functionals and data-functionals

(21)
$$\Lambda Vu = Us$$

This result has been utilized in various ways in the solution of practical inverse problems; e.g. Gilbert (1971), Wiggins (1972), Jackson (1972). For example, it can be used to identify the data-functional corresponding to a given solution-functional by first representing the given solution-functional as a linear combination of the following canonical solution-functionals.

(22)
$$v_i^T u = u_i^T s / \sigma_i$$

More importantly, when the data-functionals cannot be recovered exactly, (22) identifies the solution-functionals which the corresponding data-functionals will recover with greatest reliability (i.e. the functionals for which $\sigma_i \gg 0$) . This clarifies the point made in (ii) above abouto the role played by the improperly posedness of (12) in the linear functional strategy. Thus, there is a need to identify solution-functionals which the corresponding data-functionals will recover with some reliability (or to compute the solution-functionals directly having first identified their degree of posedness through the form of their corresponding data-functionals).

Various authors have proposed measures for the information in linear ill-posed problems; e.g. the essential dimension of Newsam (1984); and the intrinsic rank of Wahba (1980) and Lukas (1981). The above comment about the reliability of solution-functionals determined by their corresponding data-functionals can be related to such measures.

It is clear that the above discussion extends naturally to integral operators with square integrable kernels through the use of the Hilbert-Schmidt decomposition for such operators.

<u>Variational Characterizations for the Determination of the Linear Functionals</u>

Because of the advantages associated with having a variational characterization for eigenvalues, it is natural to seek variational characterizations for the determination of linear functionals such as the solution-functional $m_\theta(u)$.

A starting point is the following variational characterization for $m_\theta(u)$: _the one and only pair of functions for which the functional_

$$F(u,v) = (Ku,v) - (u,\theta) - (s,v).$$

is stationary in $(\underline{D}(K), \underline{D}(K^*))$ _is the pair_ $u = u_s$ (the solution of (12) assuming it exists), $v = \varphi$ (where φ satisfies (17)); _and for this pair the stationary value of_ $F(u,v)$ _is defined by_ $F(u_s,\varphi) = -m_\theta(u)$.

Using this result, it is not difficult to establish the following result which is often referred to as the _Schwinger–Levine Principle_ (Stakgold (1979)): _The only pair of functions for which the functional_

$$R(u,v) = \alpha\beta/\gamma \ , \ \alpha = (\theta,u), \ b = (s,v), \ \gamma = (Ku,v),$$

is stationary in $(\underline{D}(K),\underline{D}(K^*))$ _is the pair_ $u = c_1 u_s$, $v = c_2\varphi$, _where_ c_1 _and_ c_2 _are arbitrary constants, and for this pair the stationary value of_ $R(u,v)$ _is defined by_

$$R(c_1 u_s, c_2\varphi) = m_\theta(u) \ .$$

This variational formulation extends the Rayleigh quotient formulation for eigenvalues in that the use of first order accurate approximations for u and v yield second order accurate approximations for the functional $m_\theta(u)$, and these approximations will always be lower bounds for $m_\theta(u)$.

A number of authors have proposed and used such variational formulations for functionals in applications. They include Pack (1982) and Pack _et al_ (1984).

§5 CHARACTERIZATION OF THE STABILIZATION OF THE LINEAR FUNCTIONAL STRATEGY AND EXTENSIONS

Various stabilization interpretations for the linear functional stratey have already been discussed at appropriate places above. They can be summarized as follows:

(i) Limiting information determined and used for inference purposes about the solutions of improperly posed problems to bounded linear functionals defined on them, such as generalised moments, is in its own right a form of stabilization.

(ii) The degree of improperly posedness of solution-functionals can be characterized in terms of their corresponding data-functionals (when they exist); which represent the computationally efficient way in which to evaluate the solution-functionals (subject to the privisos discussed in §4). Thus, the inherent degree of smoothing in the data-functionals can be used to compare the posedness of the corresponding solution-functionals. Thus, in the foliage angle distribution problem, the functionals proposed by Suits (1972) are badly posed compared with those analysed by Anderssen $\underline{et\ al}$ (1985) since they reduce to an endpoint evaluations of the signal in the data (rather than smooth integrals of the signal).

(iii) In the linear functional strategy, the ill-posedness of the inverse problem (12) has been transferred to the transformation (17) which defines the mapping from the data-functionals to the solution-functionals. Even when the transformation (17) must be solved numerically, the solution is superior to that for the original inverse problem since θ is known analytically.

(iv) Even when the solution-functional must be computed directly using an approximation $\hat{u}$ to the solution u of (12), the Schwinger-Levine Principal (discussed in Section 4) shows that there is an essential stabilization in the determination of $m_\theta(\hat{u})$. In particular, it shows that good estimates for $m_\theta(u)$ result even when u_s (the solution of (12)) and φ are only available as approximations.

Another approach is to extend to linear functionals a formalism developed by Linz (1984) to characaterize the type of constraints under which an improperly posed problem becomes properly posed. In fact, Linz showed that if the constraint P is preserved for linear combinations of the approximate solutions u_1 and u_2 {viz. $P(u_1)$ and $P(u_2)$ imply $P(\alpha u_1 + \beta u_2)$} and there exists a $k > 0$ such that

(25) $$\|u\| = 1 \quad \text{implies} \quad \|Ku\| \geq k$$

for all u satisfying the constraint P, then if u_1 and u_2 are two solutions of (12) such that

(26) $$\|Ku_1 - s\| \leq \epsilon , \quad \|Ku_2 - s\| \leq \epsilon ,$$

then the problem of computing such solutions is properly posed in the sense that

(27) $$\|u_1 - u_2\| \leq c\epsilon , \quad c = 2/k .$$

Linz calls c the <u>condition number</u> of the properly posed problem.
In the case where P corresponds to the constraint that

(28) $$u \in \Psi \equiv \text{span}(\psi_1, \psi_2, \ldots, \psi_n) \subset \underline{\underline{H}} .$$

it follows from Linz's work that (assuming Φ and the null space of K are disjoint)

(29) $$k = (\sigma_1^{(n)})^{1/4} , \sigma_1^{(n)} = \min_{\|x\|=1} x^T (A^{-1}B)^T (A^{-1}B) x .$$

where

(30) $$A_{ij} = (\psi_i, \psi_j) , \quad B_{ij} = (K\psi_i, K\psi_j) .$$

Thus, for example, the constraining power of low parameterization (which is a popular form of informal regularization) follows from the fact that

(31) $$\sigma_1^{(n+1)} \leq \sigma_1^{(n)} , \quad n \geq 1.$$

One way of extending this work to the linear functionals $m_\theta(u)$

would be to utilize Linz's result directly by observing that , on Ψ ,

$$(32) \qquad |m_\theta(u_1-u_2)| = \left|\int_0^1 \hat{\theta}(u_1-u_2)dx\right| \leq \left\{\int_0^1 \hat{\theta}^2 dx\right\}^{1/2} \|u_1-u_2\|$$

$$\leq \left[2\left\{\int_0^1 \hat{\theta}^2 dx\right\}^{1/2}/k\right]\epsilon \ ,$$

where $\hat{\theta}$ denotes the orthogonal projection of θ onto ψ . Though informative, this estimate is rather crude.

Another way is to exploit the existence of a transformation (17) from the solution-functionals $m_\theta(u)$ to the data-functionals $m_\varphi(s)$. In fact, under the assumption that $\Psi \subset \underline{\underline{R}}(K^*)$, it follows that there exists a $\hat{\varphi}$ such that

$$(33) \qquad m_\theta(u) = (u,\hat{\theta}) = (u,K^*\hat{\varphi}) = (Ku,\hat{\varphi})$$

and hence

$$(34) \qquad |m_\theta(u_1) - m_\theta(u_2)| \leq \| K(u_1-u_2)\| \ \|\hat{\varphi}\| = 2\|\hat{\varphi}\|\epsilon \ .$$

Thus, in the sense of Linz (1984), the condition number of the properly posed problem for evaluating the solution-functionals on Ψ is $2\|\hat{\varphi}\|$. In particular, the result (34) can be taken as a proof that the evaluation of such functionals on Ψ is properly posed.

In this paper, we have not pursued the application of the linear functional strategy to identification problems for partial differential equations, such as the transmissivity problem for ground water flow (Yeh (1986)). One way in which it can be applied is through the use of the weak formulatrion for the partial differential equation involved. An application of this idea to the transmissivity problem is considred in Anderssen and Dietrich (1986).

REFERENCES

1. Anderssen, R.S.,(1980). On the use of linear funcitonals for Abel-type integral equations in applications. In: R.S. Anderssen, F.R. de Hoog and M.A. Lukas (Eds.), The Application and Numerical Solution of Integral Equations. Sijthoff and Noordhoff, Alpen aan de Rijn, pp. 195-221, 1980.

2. Anderssen, R.S. and Bloomfield, P. (1974) A time series approach to numerical differentiation, Technometrics 16 (1974), 69-75.

3. Anderssen, R.S. and Campbell, M.C.W. (1984). Computational aspects associated with the direct use of indirect measurements: refractive index of biological lenses, in Computational Techniques and Applications: CTAC-83, Eds J. Noye and C. Fletcher, North Holland, 1984, p. 893-902.

4. Anderssen, R.S. and de Hoog F.R. (1981). On the method of Chan and Lu for Abel's integral equation, J. Phys. A: Math Gen 14 (1981), 3117-3121.

5. Anderssen, R.S. de Hoog, F.R. (1982) Application and numerical solution of Abel-type integral equations, Mathematics Research Report No.7, Australian National University, 1982.

6. Anderssen, R.S. and de Hoog, F.R., 1984. A framework for studying the application of mathematics in industry. In: H. Neunzert (Ed.), Proceedings of the Conference Mathematics in Industry, October 24-28, 1983, Oberwolfach, B.G. Teubner, Stuttgart, 1984.

7. Anderseen, R.S. and Dietrich, C. (1986). The inverse problem of aquifer transmissivity identification, Proceedings of the St. Wolfgang See International Conference on Improperly and Ill-Posed Problems, Academic Press, 1986.

8. Anderssen, R.S. and Jackett, D.R., 1984. Linear functionals of foliage angle density. J.Aust. Math. Soc., Ser. B,25 (1984), 431-442.

9. Anderssen, R.S. and Jackett, D.R., (1985). Use of the linear functional strategy for assessing the status of plant canopies. In: C.M. Elliott and S. McKee (Eds.), Industrial Numerical Analysis - Case Histories, Oxford University Press, Oxford, 1985.

10.   Anderssen, R.S., Jackett, D.R. and Jupp, D.L.B., (1984). Linear functionals of the foliage angle distribution as tools to study the structure of plant canopies. _Aust. J. Bot._, 32(1984), 147-156.

11.   Anderssen, R.S., Jackett, D.R., Jupp, D.L.B. and Norman J.M. (1985). Interpretation of and simple formulas for some key linear functionals of the foliage angle distribution, _Agric. and Forest Met_ 36 (1985), 165-188.

12.   Atkinson, Jr.,A.W., John, P.C.L. and Gunning, B.E.S. (1974). The growth and division of the single mitachondrion and other organelles during the cell cycle of Chlorella, studied by quantitative stereology and three dimensional reconstruction, _Protoplasma_ 81 (1974), 77-109.

13.   Barton, N.G. and Gray, J.D. (1986). _Proceedings of the 1985 Mathematics-in-Industry Study Group,_ CSIRO Div. Math and Statistics, 1986.

14.   Bear, J. (1972). _Dynamics of Fluids in Porous Media,_ American Elsevier, New York, 1972.

15.   Buchanan, D.J. (1978). The propogation of attenuated SH channel waves _Geophys. Prosp._ 26 (1978), 16-28.

16.   Dobrin, M.B. (1960). _Introduction to Geophysical Prospecting,_ McGraw-Hill Book Co., New York, 1960.

17.   Gilbert, F. (1971). Ranking and winnowing gross Earth data for inversion and resolution, _Geophys. J.R. Astr.Soc_ 23 (1971), 125-128.

18.   Golberg, M.A. (1979). A method of adjoints for solving some ill-posed equations of the first kind, _Applied Mathematics and Computation_ 5 (1979), 123-130.

19.   Golub, G.H. and Van Loan, C.F. (1983). _Matrix Computations,_ John Hopkins University Press, Baltimore, 1983.

20.   Gordon, R., Herman, G.T. and Johnson, S.A. (1975). Image reconstruction from projections, _Scientific American_ 233 (1975), 56-68.

21.   Hyam, E.D. and Nutting, J. (1956). The tempering of plain carbon steels, _J. Iron and Steel Inst._ 148 (1956), 148-165.

22.   Koch, I. and Tarlowski, C. (1986). A new formulation of the
       magnetic relief problem. Geophysics (submitted)

23.   Jackson, D.D. (1972). Interpretation of inaccurate, insufficient
       and inconsistent data. Geophys. J.R. astr.Soc. 28 (1972), 97-109.

24.   Jakeman, A.J. and Anderssen, R.S. (1975). Abel type integral
       equations in stereology. I. General discusion. J. Micros. 105
       (1975) 121-133.

25.   Linz, P. (1984). Uncertainty in the solution of linear operator
       equations. BIT 24 (1984), 92-101.

26.   Lukas, M.A. (1981). Regularization of Linear Operator Equations,
       Ph.D. thesis, Australian Natinal University, Canberra, Australia,
       1981.

27.   McLaughlin, J.R. (1986). Analytical methods for recovering
       coeficients in differential equations from spectral data. SIAM
       Review 28 (1986), 53-72.

28.   Moran, P.A.P. (1972). The probabilistic basis of stereology,
       Special Supplement to Adv. Appl. Prob. 4 (1972), 69-91.

29.   Newsam, G.N. (1984). Measures of informatin in linear ill-posed
       problems, CMA Research Report CMA-R28-84, Australian National
       University, 1984.

30.   Pack, D.C. (1982). The development of bivariational principles for
       the calculation of upper and lower bounds, Arch. Mech. 34 (1982),
       275-285.

31.   Pack, D.C., Cole, R.J. and Mika, J. (1984). Upper and lower bounds
       of bilinear functionals in non-linear problems, IMAJ. Appl. Math.
       32 (1984), 253-266.

32.   Sabatier, P.C. (1984). Well-posed questions and exploration of the
       space of parameters in linear and non-linear inversion, Proceedings
       of a Conference on Inverse Problems of Accoustical and Elastic
       Waves, Editors F. Santora, Y.H. Pao and W.W. Symes, SIAM,
       Philadelphia, p.82-103, 1984.

33.   Stakgold, I. (1979). Green's Functions and Boundary Value Problems,
       John Wiley, New York, 1979.

34.   Suits, G.H.., (1972). The cause of azimuthal variations in directional reflectance of vegative canopies. Remote Sensing Environ. 2:(1972), 175-182.

35.   Theocaris, P.S. (1967). Moiré topography of curved surfaces, Exp. Mech. 7 (1967), 289-296.

36.   Theocaris, P.S. (1969) Moiré Fringes in Strain Analysis, Pergamon Press, London, 1969.

37.   Wahba, G. (1980). Ill-posed problems: numerical and statistical methods for midly, moderately and severely ill-posed problems with noisy data, Technical Report no. 595, Dept. of Statistics, University of Wisconsin, Madison, Wisc., 1980.

38.   Wiggins, R.A. (1972). The general linear inverse problem: implications of surface waves and free oscillations for Earth structure, Reviews of Geophysics and Space Physics 10 (1972), 251-285.

39.   Yeh, W.W.-G. (1986). Review of parameter identification procedures in ground water hydrology: the inverse problem, Water Resources Research 22 (1986), 95-108.

International Series of
Numerical Mathematics, Vol. 77
© 1986 Birkhäuser Verlag Basel

Determination of a Source Term in a Linear Parabolic Differential Equation with Mixed Boundary Conditions

John R. Cannon and Yanping Lin

Abstract: We consider the problem of determining $u = u(x,t)$
and $f = f(x)$ which satisfy $Lu = (p(x)u_x)_x - q(x)u - \rho(x)u_t = \rho(x)f(x)$, $0 < x < 1$, $0 < t \le T$, $u(x,0) = \varphi(x)$, $u(0,t) = \sigma_1(t)$,
and $u_x(1,t) + \beta u(1,t) = \sigma_2(t)$ from the additional specification of $p(0)u_x(0,t) = g(t)$. The problem is not well posed
in the sense of Hadamard. Uniqueness is demonstrated. Continuous
dependence of u and f upon the data is demonstrated for f twice
continously differentiable in $0 \le x \le 1$ with $f(0) = 0$ and
$f'(1) + \beta f(1) = 0$ and with first and second derivatives bounded
in absolute value by the known positive constant K. The continuous dependence is shown to be asymptotically logarithmic.

I. Introduction

In this paper we generalized a result of one of the authors [6,10] for

heat equation from first boundary data to mixed boundary data. The problem

(1.1) represents a generalization of the work of Cannon [6] and a

modification of the work of Cannon and Ewing [10]. Suppose that an unknown

heat source f, which is dependent only upon the spatial variable, is

operating on the unit interval. Suppose that the initial and mixed

boundary data are overspecified by the additional measurement of the

heat flow rate at the boundary $x = 0$ over an interval of time

$0 \le \delta \le t \le T$, $T > 0$.

The mathematical problem can be stated as: find the pair $u = u(x,t)$ and $f = f(x)$ that satisfy

$$(1.1a) \qquad Lu = (p(x)u_x)_x - q(x)u - \rho(x)u_t = \rho(x)f(x), \qquad 0 < x < 1, \ 0 < t \leq T,$$

$$(1.1b) \qquad u(x,0) = \varphi(x), \qquad 0 \leq x \leq 1,$$

$$(1.1c) \qquad u(0,t) = \sigma_1(t), \qquad 0 < t \leq T,$$

$$(1.1d) \qquad u_x(1,t) + \beta u(1,t) = \sigma_2(t), \qquad 0 < t \leq T,$$

$$(1.1e) \qquad p(0)\, u_x(0,t) = g(t), \qquad 0 \leq \delta \leq t \leq T,$$

where φ, σ_1, σ_2 and g are known functions. We note that the form of of the equation (1.1a) can be achieved from the form of a general equation

$$(1.2) \qquad C(x)\, U_{xx} + E(x) + F(x)U - U_t = f(x)$$

by the use of the integrating factor $\rho(x)$ as in Petrovskii [13].

In order to demonstrate uniqueness, let (u_i, f_i), $i = 1,2$, denote two solutions to (1.1). Setting $w = u_1 - u_2$ and $\tilde{F} = f_1 - f_2$, it follows that

$$(1.3a) \qquad Lw = \tilde{F}, \qquad 0 < x < 1, \ 0 < t \leq T,$$

$$(1.3b) \qquad w(x,0) = 0, \qquad 0 \leq x \leq 1,$$

$$(1.3c) \qquad w(0,t) = 0, \qquad 0 < t \leq T,$$

$$(1.3d) \qquad w_x(1,t) + \beta w(1,t) = 0, \qquad 0 < t \leq T,$$

$$(1.3e) \qquad w_x(0,t) = 0, \qquad \delta \leq t \leq T.$$

By standard separation of variables, we obtain

$$(1.4) \qquad w(x,t) = \sum_{n=1}^{\infty} (-1)C_n(\lambda_n^{-1} - \lambda_n^{-1}\exp\{-\lambda_n t\})\, X_n(x),$$

where λ_n and $X_n(x)$, $n = 1,2,\ldots$, are the eigenvalues and orthogonal eigenfunctions, respectively, for the Sturm–Liouville problem

$$(1.5) \qquad (p(x)X_n')' + (\lambda_n \rho(x) - q(x))X_n = 0, \qquad\qquad 0 < x < 1,$$

and

$$(1.6) \qquad \begin{aligned} & X_n'(0) = 0, \\ & X_n'(1) + \beta\, X_n(1) = 0 , \end{aligned}$$

and where

$$(1.7) \qquad C_n = \int_0^1 \rho \tilde{F} X_n \, dx / \int_0^1 \rho X_n^2 \, dx, \qquad n = 1,2,\dots \;\;.$$

From (1.3e) and next sections, we know that

$$(1.8) \qquad 0 = \sum_{n=1}^{\infty} (-1)\, C_n\, (\lambda_n^{-1} - \lambda_n^{-1}\, \exp\{-\lambda_n t\}) X_n'(0)$$

which is an analytic function in $\mathrm{Re}\, t \geq 0$ with the form

$$(1.9) \qquad b_0 + \sum_{n=1}^{\infty} b_n \, \exp\{-\lambda_n t\}.$$

By (1.8) it follows that

$$(1.10) \qquad C_n \lambda_n^{-1} X_n'(0) = 0, \qquad n = 1,2,\dots,$$

since $X_n'(0) \neq 0$, $n = 1,2,\dots$, we see that

$$(1.11) \qquad C_n = 0, \qquad n = 1,2,\dots,$$

and that $f_1 \equiv f_2$ and $u_1 \equiv u_2$.

Now we consider the problem

$$(1.12a) \qquad \Psi_{xx} - \Psi_t = f(x), \qquad\qquad 0 < x < 1, \qquad 0 < t \leq T,$$

$$(1.12b) \qquad \Psi(x,0) = 0, \qquad\qquad 0 \leq x \leq 1,$$

$$(1.12c) \qquad \Psi(0,t) = \Psi_x(1,t) = 0, \qquad\qquad 0 < t \leq T,$$

$$(1.12d) \qquad \Psi_x(0,t) = (-1)\, a_n (n + \tfrac{1}{2})^{-1} \pi^{-1} (1 - \exp\{-(n+\tfrac{1}{2})^2 \pi^2 t\}), \quad 0 < t \leq T.$$

For each positive n, the solution (Ψ_n, f_n) is

$$(1.13) \qquad f_n(x) = a_n \sin(n + \tfrac{1}{2})\pi x,$$

and

$$(1.14) \qquad \Psi_n(x,t) = (-1)\, a_n (n + \tfrac{1}{2})^{-2} \pi^{-2} (1 - \exp\{-(n+\tfrac{1}{2})^2 \pi^2 t\}) \sin(n+\tfrac{1}{2})\pi x.$$

First, choosing $a_n = \sqrt{n}$, a sequence of solutions f_n is obtained whose norms tend to infinity while the data $\Psi_x(0,t)$ tends to zero uniformly; next choosing $a_n = 1$, a sequence of solutions is obtained whose norms remain constant while the data $\Psi_x(0,t)$ tends to zero uniformly. Consequently, it follows that the problem (1.1) is not well-posed in the sense of Hadamard. Hence, a priori information concerning f is necessary in order to insure continuous dependence of the solution (u,f) upon the data φ, σ_1, σ_2 and g.

The solution (u,f) can be written in the form $(u,f) = (v,0) + (z,f)$, where v satisfies

$$(1.15a) \quad Lv = 0, \qquad\qquad 0 < x < 1, \quad 0 < t \leq T,$$

$$(1.15b) \quad v(x,0) = \varphi(x), \qquad\qquad 0 \leq x \leq 1,$$

$$(1.15c) \quad v(0,t) = \sigma_1(t), \qquad\qquad 0 < t \leq T,$$

$$(1.15d) \quad v_x(1,t) + \beta v(1,t) = \sigma_2(t), \qquad\qquad 0 < t \leq T$$

where (z,f) satisfies

$$(1.16a) \quad Lz = \rho f, \qquad\qquad 0 < x < 1, \quad 0 < t \leq T,$$

$$(1.16b) \quad z(x,0) = 0, \qquad\qquad 0 \leq x \leq 1,$$

$$(1.16c) \quad z(0,t) = 0, \qquad\qquad 0 < t \leq T,$$

$$(1.16d) \quad z_x(1,t) + \beta z(1,t) = 0, \qquad\qquad 0 < t \leq T,$$

$$(1.16e) \quad p(0)z_x(0,t) = G(t), \qquad\qquad 0 \leq t \leq T,$$

and where

$$G(t) = g(t) - \rho(0)\, v_x(0,t).$$

From Friedman [11] it is easy to see that v and v_x depend continuously upon the data φ, σ_1, σ_2, σ_1' and σ_2'. It will suffice to show that (z,f) depends continuously upon G. Hence, we shall restrict our attention to (1.16) and its solution (z,f).

35

It is convenient here to list some hypotheses that will be needed below.

<u>Assumption 1</u>. We shall assume that f is twice continuously differentiable in $0 \le x \le 1$, that $f(0) = 0$, $f'(1) + \beta f(1) = 0$, and that there exists a positive constant K, such that f and its first and second derivatives with respect to x, f' and f", are uniformly bounded in absolute value by K.

<u>Assumption 2</u>. The functions p, q and ρ are uniformly Hölder continuously in $0 \le x \le 1$ and satisfy

(1.17a) $\qquad 0 < p_* \le p(x) \le p^*,$

(1.17b) $\qquad 0 < \rho_* \le \rho(x) \le \rho^*,$

(1.17c) $\qquad 0 \le q_* \le q(x) \le q^*,$

and

(1.17d) $\qquad |\rho'(x)| \le p'^*,$

moreover, $(p\rho)"$ is assumed to be a continuous function.

<u>Remark</u>. We shall refer to the constants in (1.17) either individually or collectively through the symbol D which denotes the set of constants in (1.17). Also in what follows, we shall let

(1.18) $\qquad \|f\|_{[a,b]} = \sup_{a \le \zeta \le b} |f(\zeta)|$

for any real valued function f defined on $a \le \zeta \le b$, and shall let

(1.19) $\qquad \|f\|_2 = (\int_0^1 [f(x)]^2 dx)^{\frac{1}{2}}$

for any real valued function $f \in L^2(0,1)$.

Now we state the main theorem as the following.

<u>Theorem</u>: If assumptions 1 and 2 hold, then there exist positive valued sequences $\{A_N^{(i)}\}$ and $\{B_N^{(i)}\}$, $i = 1,2$, such that

$$\lim_{N \to \infty} A_N^{(i)} = \infty \quad \text{and} \quad \lim_{N \to \infty} B_N^{(i)} = 0, \quad i = 1,2,$$

and there exists a positive constant ν, $0 < \nu < 1$, such that for each integer $N > 0$,

$$(1.20) \qquad \|f\|_2^2 \leq A_N^{(1)} \eta^\nu + B_N^{(1)} ,$$

and

$$(1.21) \qquad \|z(\cdot,t)\|_2^2 \leq A_N^{(2)} \eta^\nu + B_N^{(2)} ,$$

where

$$(1.22) \qquad \eta = \|G\|_{[\delta,T]}$$

and (z,f) is the solution of (1.16).

II. <u>Outline of the Proof of the Theorem</u>

The assumed smoothness of f allows us to expand it into the series form

$$(2.1) \qquad f(x) = \sum_{n=1}^{\infty} C_n X_n(x)$$

where the X_n are the orthogonal eigenfunctions corresponding respectively to the eigenvalues λ_n for the Sturm-Liouville problem (1.5) - (1.6) and C_n are defined in (1.7) with $\tilde{F}$ replaced by f which are the Fourier coefficients of f. We shall note here that the λ_n are positive and form a monotone increasing sequence such that $\lim_{n \to \infty} \lambda_n = \infty$.

An estimation of f can be obtained in terms of η once the X_n are bounded in terms of D and C_n are estimated in terms of η.

As in section I, separation of variables yields the formal representation

$$(2.2) \quad z(x,t) = \sum_{n=1}^{\infty} (-1) \, C_n \, [\lambda_n^{-1} - \lambda_n^{-1} \exp\{-\lambda_n t\}] X_n(x)$$

for z in terms of the Fourier coefficients of f.

In section III, we shall show that the X_n are uniformly bounded by some fixed positive constant, that the $X_n'(x)$ are bounded in terms of $\lambda_n^{\frac{1}{2}}$, that $\lambda_n = O(n^2)$, and that $|C_n| = O(\lambda_n^{-1})$ $(n \to \infty)$. Consequently, the formal representation for z converges absolutely and uniformly and its partial derivatives with respect to x can be obtained by differentiating term by term, we see that

$$(2.3) \quad F(t) = p(0) \, z_x(0,t)$$
$$= \sum_{n=1}^{\infty} C_n \, \lambda_n^{-1} \, p(0) \, X_n'(0) \, (\exp\{-\lambda_n t\}-1).$$

it follows that $F(t)$ is a bounded analytic function in $\mathrm{Re}\ t \geq 0$. For $t^* \in (\delta, T)$, we have

$$(2.4) \quad F'(t^* + i\tau) = \sum_{n=1}^{\infty} A_n \, \exp\{-\lambda_n \tau i\}$$

where

$$(2.5) \quad A_n = (-1) \, C_n p(0) \, X_n'(0) \, \exp\{-\lambda_n t^*\}, \quad n = 1,2,\ldots .$$

Binmore [2] has shown that the A_n's can be bounded in terms of $\sup_{|\tau| \leq \tau_n} |F'(t^* + i\tau)|$, where

$$(2.6) \quad \tau_n = \frac{\pi}{2} \, (n\lambda_n^{-1} + \sum_{j=n+1}^{\infty} \lambda_j^{-1}), \quad n = 1,2,\ldots .$$

By using Lindelöf and Carleman type estimates for analytic functions [7], the quantity $\sup_{|\tau| \leq \tau_n} |F'(t^* + it)|$ can be bounded in terms of η.

Thus, A_n's can be bounded in terms of η. The estimates of C_n follows once lower bounds for $X_n'(0)$ are obtained.

The lower bounds for $X_n'(0)$ and the relevant estimates for the solution of the Sturm-Liouville problem are derived from the known asymptotic results in section III. In section IV the analytic function theory results of Carleman, Lindelöf and Binmore are recalled. The proof of the theorem is given in section V by referring to the results presented in sections III and IV.

III.   Some Estimates of the Sturm-Liouville Problem.

First we state the following result.

Theorem 3.1. Consider the Sturm-Liouville problem

(3.1) $y'' + (\lambda - Q)y = 0,$ $0 < x < 1,$

and

(3.2)
$$y(0) \cos \alpha + y'(0) \sin \alpha = 0,$$
$$y(1) \cos \theta + y'(1) \sin \theta = 0$$

where $Q = Q(x)$, $0 \le x \le 1$, is a continuous function. Then we have

(A) The eigenvalues λ_k of (3.1)- (3.2) satisfy

(3.3) $\sqrt{\lambda_k} = k\pi + O(k^{-1}),$ $(k \to \infty)$

for $\sin\alpha \ne 0$, $\sin\theta \ne 0$ and $\sin\alpha = 0$, $\sin\theta = 0$, and

(3.4) $\sqrt{\lambda_k} = (k + \tfrac{1}{2})\pi + O(k^{-1}),$ $(k \to \infty)$

for $\sin\alpha = 0$, $\sin\theta \ne 0$ and $\sin\alpha \ne 0$, $\sin\theta = 0$.

(B) If $y(x,\lambda)$ satisfies

(3.5) $y(1) = 1,$ $y'(1) = H,$

39

then

$$(3.6) \qquad y(x) = \cos\sqrt{\pi}(1-x) - H\pi^{-\frac{1}{2}} \sin\sqrt{\lambda}(1-x)$$

$$-\frac{1}{\sqrt{\lambda}} \int_x^1 \sin\sqrt{\lambda}(x-\tau)Q(\tau)y(\tau)d\tau$$

(C) If the eigenfunctions y_k corresponding to the eigenvalues λ_k satisfy (3.5), then we have

$$(3.7) \qquad \int_0^1 y_k^2(x)dx = 2^{-1} + 0(k^{-2}), \qquad (k \to \infty).$$

Proof. See [12, 14].

Now, we make the following transformations.

$$(3.8) \qquad \left\{ \begin{array}{l} \omega = \dfrac{1}{\bar{K}} \int_0^x \dfrac{\rho(x)}{p(x)}^{\frac{1}{2}} dx \\[3ex] \bar{K} = \int_0^1 \dfrac{\rho(x)}{p(x)}^{\frac{1}{2}} dx \\[3ex] Y_n(\omega) = (\rho(x)p(x))^{\frac{1}{4}} X_n(x) \\[3ex] \mu_n^2 = \bar{K}^2 \lambda_n \end{array} \right.$$

Then (1.5) - (1.6) becomes

$$(3.9) \qquad Y_n'' + (\mu_n^2 - \bar{g}(\omega))Y_n = 0, \qquad 0 < \omega < 1,$$

and

$$(3.10) \qquad \left\{ \begin{array}{l} Y_n(0) = 0, \\[1ex] Y_n'(1) + \bar{\beta}\,\bar{Y}_n(1) = 0, \end{array} \right.$$

where

$$(3.11) \qquad \bar{\beta} = \bar{K} \left[\frac{\beta - \frac{1}{2}(pp' + \rho p')}{p^{\frac{1}{2}}} \right]_{x=1} \quad ,$$

and

$$\bar{g}(\omega) = \bar{f}'(\omega)/\bar{f}(\omega) - \bar{K}^2 \bar{h}(\omega) \ ,$$

$$(3.12) \qquad \bar{f}(\omega) = (\rho(x)p(x))^{\frac{1}{4}}$$

$$\bar{h}(\omega) = q(x)/p(x) \ .$$

In order to apply Theorem 3.1, we normalize the eigenfunctions $X_n(x)$ in the following way:

$$(3.13) \qquad X_n(1)(p(1)\rho(1))^{\frac{1}{4}} = 1, \qquad n = 1,2,\ldots \ .$$

Note that $X_n(1) \neq 0$ since otherwise X_n would be identically zero.

From (3.8) and (3.13), we see that

$$(3.14) \qquad Y_n(1) = 1, \qquad Y'(1) = -\bar{\beta} \ .$$

From application of Theorem 3.1 with (3.14), it follows that

$$(3.15) \qquad \int_0^1 Y_n^2 \, d\omega = \frac{1}{2} + O(n^{-2}) \ , \ (n \to \infty).$$

By (3.8) and (3.15), it is easy to see that

$$(3.16) \qquad \int_0^1 \rho X_n^2 dx = \frac{\bar{K}}{2} + O(n^{-2}), \qquad (n \to \infty).$$

Now we consider the upper bounds for $X_n'(0)$, $n = 1,2,\ldots$. From (3.6) we see that

$$(3.17) \qquad Y_n(\omega) = \cos(\mu_n(1-\omega)) - \frac{\bar{\beta}}{\mu_1} \sin(\mu_n(1-\omega))$$

$$+ \ \mu_n^{-1} \int_0^\omega \sin\{\mu_n(\omega-\tau)\}\bar{g}(\tau)Y_n(\tau)d\tau$$

By Gronwall's Lemma it is easy to see that

(3.18) $\quad |Y_n(\omega)| \leq (1 + \frac{|\bar{\beta}|}{\mu_1}) \exp\{\frac{1}{\mu_1} \int_0^1 |\bar{g}(\omega)| d\omega\}$.

Hence by (3.8), it follows that there exists a positive constant $K_1 = K_1(D,\beta)$ such that

(3.19) $\qquad |X_n(x)| \leq K_1, \quad n = 1,2,\ldots , \quad 0 \leq x \leq 1.$

Differentiating (3.17) with respect to ω and using (3.18) and observing that $\mu_n > 0$, $n = 1,2,\ldots$, and $\mu_n \to \infty$ $(n \to \infty)$, we know that there exists a positive constant $K_2 = K_2(D,\beta)$ such that

(3.20) $\qquad |Y_n'(\omega)| \leq K_2\mu_n$.

From (3.8) it follows that

(3.21) $\quad Y_n'(\omega) \frac{d\omega}{dx} = [(p\rho)^{\frac{1}{4}}]' X_n(x) + (p\rho)^{\frac{1}{4}}X_n'(x).$

Combining (3.19), (3.20) and (3.21), we see that there exists a positive constant $K_3 = K_3(D,\beta)$ such that

(3.22) $\quad |X_n'| \leq K_3 \lambda_n^{\frac{1}{4}}, n = 1,2,\ldots$.

Multiplying equation (1.5) by $f(x)$ and integrating from 0 to 1, we obtain

(3.23) $\quad \lambda_n \int_0^1 \rho f X_n dx = \int_0^1 fq X_n dx - \int_0^1 f(pX_n')'dx.$

By assumption 1 and integrating by parts together with (3.19) it follows that there exists a positive constant $K_4 = K_4(D,\beta,K)$ such that

(3.24) $\quad |\int_0^1 \rho f X_n dx| \leq K_4/\lambda_n, \quad n = 1,2,\ldots,.$

Hence from (3.16) there exists a positive constant $K_5 = K_5(D,\beta,K)$ such that

(3.25) $\quad |C_n| = | \int_0^1 f\rho X_n dx / \int_0^1 \rho X_n^2 dx| \leq K_5/\lambda_n, \quad n = 1,2,\ldots$.

In section II we stated that we need lower bounds of $X_n'(0)$ to get estimates of C_n's in terms of η. Now let (V,ψ) denote the solution of

(3.26) $\quad \begin{cases} V' = \psi, & 0 < x < 1, \\ \psi' = -p'(x) [p(x)]^{-1}\psi - (\lambda_n \rho(x) - q(x))[p(x)]^{-1}V, & 0 < x < 1, \end{cases}$

and

$$(3.27) \quad \begin{cases} V(0) = 0, \\ \psi(0) = 1, \end{cases}$$

clearly, $V = X_n(x)/X_n'(0)$ is the solution and

$$(3.28) \qquad \int_0^1 \rho V^2 dx = \int_0^1 \rho X_n^2 dx / |X_n'(0)|^2 .$$

By (3.16) we see that

$$(3.29) \qquad |X_n'(0)|^2 = (\frac{\bar{K}}{2} + O(n^{-2}))(\int_0^1 \rho V^2(x)dx)^{-1}, \quad (n \to \infty)$$

so that there exists a positive $K_6 = K_6(D,\beta)$ such that

$$(3.30) \quad |X_n'(0)|^2 \geq (\frac{\bar{K}}{2} - K_6 n^{-2}) (\int_0^1 \rho V^2 dx)^{-1}, \quad (n \to \infty).$$

Consequently, a lower estimate can be obtained from an upper estimate
of V. Set

$$(3.31) \qquad W(x) = \max \{\|V\|_{[0,x]}, \|\psi\|_{[0,x]}\} .$$

Then, it follows that there exists a constant $K_7 = K_7(D,\beta,n)$ such
that

$$(3.32) \qquad W(x) \leq 1 + K_7 \int_0^x W(\zeta)d\zeta .$$

An application of Gronwall's Lemma yields $W(x) \leq \exp\{K_7\}$ and

$$(3.33) \qquad \|V\|_{[0,1]} \leq \exp \{K_7\}.$$

Hence (3.30) and (3.33) yields for n sufficiently large

$$(3.34) \quad |X_n'(0)| \geq K_6 n^{-2} p*^{-\frac{1}{2}} \exp\{-K_7\}.$$

As $|X_n'(0)| > 0$ for all n, there exists a positive constant
$K_8 = K_8(D,\beta)$ such that for all n

$$(3.35) \qquad |X_n'(0)| \geq K_8^{-1} \exp \{-K_7\}.$$

43

<u>Remark</u>. So far we have stated and derived all of the estimates of the Sturm-Liouville problem which we shall need in section V. In fact, the upper bounds of C_n's and $X_n'(0)$ are needed to show that $F(t)$ is a bounded analytic function in Ret ≥ 0 and the lower bounds of $X_n'(0)$ will help to obtain the estimates of C_n's in terms of η.

IV. <u>Some Analytic Function Estimates</u>.

Recall in (2.4) that

$$(4.1) \qquad F'(t* + i\tau) = \sum_{n=1}^{\infty} A_n \exp\{-\lambda_n \tau i\}.$$

Binmore [2] has shown the following

<u>Theorem 4.1</u>. Suppose

$$(4.2) \qquad f(t) = \sum_{n=0}^{\infty} A_n \exp\{-\lambda_n it\}$$

is a bounded function and $0 < \lambda_0 < \lambda_1 < ... < \lambda_n < ...$ with $\sum_{n=0}^{\infty} \frac{1}{\lambda_n} < \infty$. Then

$$(4.3) \qquad |A_n| \leq \prod_{j=0}^{n-1} \cos(\frac{\pi}{2} \lambda_j \lambda_n^{-1}) \prod_{j=n+1}^{\infty} \cos(\frac{\pi}{2} \lambda_n \lambda_j^{-1}) \sup_{|\tau| \leq \tau_n} |f(\tau)|,$$

where

$$(4.4) \qquad \tau_n = \frac{\pi}{2} [\frac{n}{\lambda_n} + \sum_{j=n+1}^{\infty} \frac{1}{\lambda_j}] , \quad n = 1,2,... .$$

Proof. See [2].

For bounded analytic functions we have

<u>Theorem 4.2</u>. Let $f(z)$ be an analytic function in the rectangle

$$(4.5) \qquad \Omega = \{z = x + iy \mid a < x < b, \text{ and } c < y < d\},$$

which is continuous in the closure $\bar{\Omega}$ of Ω, for $0 < \epsilon < C$, suppose that

$$(4.6) \qquad |f(x + ic)| \leq \epsilon , \qquad a < x < b,$$

and that $|f| \leq C$ on the remaining three sides of Ω. Then for $z \in \bar{\Omega}$

$$(4.7) \qquad |f(z)| \leq C^{1-h(z)} \epsilon^{h(z)}$$

where

$$(4.8) \qquad h(z) = s(y) \sin\left(\frac{x-a}{b-a}\right)\pi,$$

and
$$(4.9) \qquad s(y) = \frac{\sinh\left[(y-d)/(b-a)\right]}{\sinh\left[(c-d)/(b-a)\right]} .$$

Proof. See [14].

In order to use Theorem 4.2, we need the following lemma.

Lemma. The τ_n, $n = 1,2,\ldots$, are uniformly bounded where τ_n is defined in (4.4).

Proof: From Theorem 3.1, we see that

$$(4.10) \qquad \lambda_n = \mu_n^2/\bar{K}^2, \qquad n = 1, 2,\ldots,$$

and

$$(4.11) \qquad \mu_n = \begin{cases} n\pi + 0(n^{-1}), \text{ or} \\ (n+\tfrac{1}{2})\pi + 0(n^{-1}), \qquad (n \to \infty) . \end{cases}$$

Combining (4.10) and (4.11) it follows that

$$(4.12) \qquad \sum_{n=1}^{\infty} \frac{1}{\lambda_n} < \infty$$

and that

$$(4.13) \qquad n/\lambda_n = 0(n^{-1}), \qquad (n \to \infty) .$$

Hence, (4.12) and (4.13) complete the lemma.

V. **Proof of the Theorem.**

Recall that

45

(5.1) $\quad F(t) = \sum_{n=1}^{\infty} C_n \lambda_n^{-1} p(0) X_n'(0) \, (\exp\{-\lambda_n t\} - 1)$

and

(5.2)
$$\begin{cases} |C_n| \le K_5/\lambda_n \,, & n = 1,2,\ldots, \\[4pt] |X_n| \le K_1, & n = 1,2,\ldots, \\[4pt] |X_n'| \le K_3 \, \lambda_n^{\frac{1}{2}}, & n = 1,2,\ldots, \\[4pt] \lambda_n = O(n^2), & (n \to \infty) \,. \end{cases}$$

Hence, it is easy to see that $F(t)$ is a bounded analytic function in the complex domain $\mathrm{Re}\, t \ge 0$, and by (2.3), it follows that

(5.3) $\qquad\qquad |F(t)| \le \eta \qquad\qquad 0 \le \delta \le t \le T,$

where η is defined by (1.22).

Moreover, there exists a positive constant $K_9 = K_9(D,\beta,K)$ such that for all $\zeta = t + i\tau$ in $\mathrm{Re}\,\zeta \ge 0$

(5.4) $\qquad\qquad |F(\zeta)| \le K_9$

where we take K_9 to exceed η.

In order to apply Theorem 4.2 to our problem, let

(5.5) $\qquad c = 0, \quad d = t^* > 0, \quad a = \delta, \quad b = T, \text{ if } \eta \ge 1,$

$\qquad\qquad c = 0, \quad d = 2\tau^* > 0, \quad a = \delta, \quad b = T, \text{ if } 0 < \eta < 1.$

As $F(\zeta)$ is a bounded analytic function $\mathrm{Re}\,\zeta \ge 0$, it follows from (5.3) – (5.4) and Theorem 4.2 that there exists a positive constant $K_{10} = K_{10}(D,\beta,K)$ such that

(5.6) $\qquad |F(\zeta)| \le K_{10}\eta^{h(\zeta)} \,.$

Since

(5.7) $\quad s'(y) = \dfrac{\pi}{b-a} \; \dfrac{\cosh\,[\pi(y-d)/(b-a)]}{\sinh\,[\pi(c-d)/(b-a)]} \le 0, \qquad c < y < d,$

and $s(0) = 1$, we take

$$(5.8) \quad \begin{cases} t^* = \left(\dfrac{T + 5\delta}{12}\right), & \delta < t^* < T, \\[2mm] \alpha = \sin\left(\dfrac{t^*-\delta}{T-\delta}\right)\pi, & \text{if } \eta \geq 1, \\[2mm] \alpha = s(\tau^*)\sin\left(\dfrac{t^*-\delta}{T-\delta}\right)\pi, & \text{if } 0 < \eta < 1 \end{cases}$$

and note that $0 < \alpha < \tfrac{1}{2}$, and from (5.6)

$$(5.9) \quad F(t^* + i\tau)| \leq K_{10}\eta^\alpha, \qquad 0 \leq \tau \leq \tau^*.$$

Similarly, we apply Theorem 4.2 to the analytic function $\overline{F(\overline{\zeta})}$

Thus there exists a positive constant $K_{11} = K_{11}(D,\beta,K)$ such that

$$(5.10) \quad |F(t^* - i\tau)| \leq K_{11}\eta^\alpha, \qquad 0 \leq \tau \leq \tau^*.$$

Hence, let

$$(5.11) \quad K_{12} = \max\{K_{10}, K_{11}\}$$

and combine (5.9) and (5.10) to obtain

$$(5.12) \quad |F(t^* + i\tau)| \leq K_{12}\eta^\alpha, \qquad |\tau| \leq \tau^*,$$

where we have taken τ^* to exceed all of τ_n $n = 1,2,\ldots$.

An elementary estimation of the Cauchy-Riemann representation formula

for $F'(\zeta)$ yields

$$(5.13) \quad |F'(t^* + i\tau)| \leq K_{13}\eta^\alpha \qquad |\tau| \leq \tau^*$$

where $K_{13} = K_{13}(D, \beta, K, \tau^*, t^*)$. From (5.1) we see that

$$(5.14) \quad F'(t^* + i\tau) = \sum_{n=1}^{\infty} A_n \exp\{-\lambda_n \tau i\},$$

where

$$(5.15) \quad A_n = (-1) C_n p(0)X_n'(0) \exp\{-\lambda_n t^*\}.$$

A straightforward application of Theorem 4.1 yields

$$(5.16) \quad |C_n p(0)X_n'(0) \exp\{-\lambda_n t^*\}| \leq H(\lambda_n)K_{13}\eta^\alpha$$

where

$$(5.17) \quad H(\lambda_n) = \prod_{j=1}^{n-1} \cos\left(\frac{\pi}{2}\lambda_j\lambda_n^{-1}\right) \prod_{j=n+1}^{\infty} \cos\left(\frac{\pi}{2}\lambda_n\lambda_j^{-1}\right), \quad n = 1,2,\ldots,$$

From (3.35), it follows that

(5.18) $\quad |C_n| \le p_* H(\lambda_n) K_8 \exp\{-\lambda_n t^* + K_7\} K_{13} n^\alpha \quad n = 1,2,\ldots,$

which we proceed to use in estimating the L^2-norm of f.

From (2.1) and the orthogonality of the X_n, we have

(5.19) $\quad \int_0^1 \rho f^2 dx = \sum_{n=1}^\infty c_n^2 \int_0^1 \rho X_n^2 dx$

which means

(5.20) $\quad \sum_{n=1}^\infty c_n^2 \int_0^1 \rho X_n^2 dx < \infty.$

From (3.16) there exists a positive constant $K_{14} = K_{14}(D,\beta)$ such that

(5.21) $\quad \int_0^1 \rho X_n^2 dx \le K_{14} \qquad\qquad n = 1,2,\ldots\ .$

Combining (5.19) and (5.21) for any $N > 0$,

(5.22) $\quad \|f\|_2^2 \le \rho_*^{-1} K_{14} \sum_{n=1}^N c_n^2 + \rho_*^{-1} \sum_{n=N+1}^\infty c_n^2 \int_0^1 \rho X_n^2 dx\ .$

By (5.18), it follows from (5.20) and (5.22) that for each $N > 0$

(5.23) $\quad \|f\|_2^2 \le A_N^{(1)} \eta^\nu + B_N^{(1)}$

where $\nu = 2\alpha$, $0 < \nu < 1$, and $A_N^{(1)}$ and $B_N^{(1)}$ are

positive and such that $\lim\limits_{N\to\infty} A_N^{(1)} = \infty$ and

$\lim\limits_{N\to\infty} B_N^{(1)} = 0.$ We see from (2.2) there exists a positive

constant $K_{15} = K_{15}(D,\beta,K)$ such that

(5.24) $\quad \|z(\cdot,t)\|_2^2 \le K_{15}\|f\|_2^2.$

Hence, the demonstration of the theorem is complete.

<u>Corollary</u>. If assumptions 1 and 2 hold, then the L^2-norm of f the

component of the solution (z,f) of (1.6) depends logarithmically upon the

uniform norm of the data G.

Proof. By theorem [9, section III] we need only to show that there exists

a minimum positive separation between the eigenvalues of the Sturm-

Liouville problem (1.5) - (1.6). In fact, from theorem 3.1 and and

$\mu_n^2 = K^{-2}\lambda_n$, we see that

48

$$(5.25) \qquad \lambda_n = \begin{cases} \dfrac{n^2\pi^2}{\bar{K}^2} + O(1), & \text{or} \\[3ex] \dfrac{(n+\frac{1}{2})^2\pi^2}{\bar{K}^2} + O(1), & (n \to \infty) \end{cases}$$

and since $\{\lambda_n\}$ is a monotone sequence, it follows that there exists a $\delta^* > 0$ that

$$(5.26) \qquad \lambda_{n+1} - \lambda_n \geq \delta^* > 0, \qquad\qquad n = 1,2,\ldots .$$

This completes the proof of the corollary.

<u>Remark</u> <u>on</u> <u>Numerical</u> <u>Procedures</u>:

Logarithmic dependence upon the data implies that any numerical method for (1.6) would be marginal at best. Any formulation of a numerical procedure for (1.6) would involve a mathematical programming problem of the least squares or linear variety. For examples of these types of formulations, we refer the reader to [6,8].

References

[1]   J.M. Anderson and K.G. Binmore, "Coefficients estimates for Lacunary
        power series and Dirichlt Series II", Proc. Lond. Math. Soc
        3(1963), pp.49-68.
[2]   K.G. Binmore, "A trigonometric inequality", J. Lond. Math. Soc.
        41(1966), pp. 693-696.
[3]   J.R. Cannon, "A priori estimate for continuation of the solution of
        the heat equation in the space variable", Annali di Mathematica
        Pura ed Applicata (IV), Vol. LXV, (1964) pp. 377-388.
[4]   J.R. Cannon, "Error estimates for some unstable continuation
        problems", J. of SIAM 12(1964), pp. 270-284.
[5]   J.R. Cannon, "A Cauchy problem for the heat equation", Annali di
        mathematica Pura ed Applicata (IV) VOl LXVI (1964) pp. 155-165.
[6]   J.R. Cannon, "Determination of an unknown heat source from over
        specified boundary data", SIAM J. Numer. Anal. 5(1968)
        pp.275-286.
[7]   J.R. Cannon, The One-Dimensional Heat Equation Encyclopedia of
        Mathematics, Addison-Wesley, Reading, MA, 1984.
[8]   J.R. Cannon and Jim Douglas, Jr., "The approximation of harmonic
        and parabolic functions on half-space from interior data",
        Centro Internazionale Mathematico Estivo, Numerical Analysis
        of Partial Differential Equations (Ispra (Varese) Italy; 1969)
        Editore Cremonese Roma,1969.
[9]   J.R. Cannon and R.E. Ewing, A direct numerical procedure for the
        cauchy problem for the heat equation, J. Math. Anal. Appl.
        56(1976), pp. 7-17.
[10]   J.R. Cannon and R.E. Ewing, Determination of a source term in a
        Linear parabolic partial differential equation. J. Appl.
        Math. and Physics, 27(1976), 393-401.
[11]   A. Friedman, Partial Differential Equation of Parabolic Type,
        Prentice Hall, Englewood Cliffs, N.J. (1964), p.146.
[12]   B.M. Leviton and I.S. Sargsjan, Introduction to Spectral Theory,
        Transactions of Math Monograph, No. 39, Amer. Math. Soc.,
        Providence, R.I. 1975, pp. 5-13.
[13]   I.G. Petrovskii, Partial Differential Equations, W. B. Saunders Co.
        1967.
[14]   Kösaku Yosida, Lecture on Differential and Integral Equations,
        Interscience publishirs Inc. N.Y. (1960) pp. 110-114.

John R. Cannon   Yanping Lin
Department of Mathematics   Department of Mathematics
Washington State University   Washington State University
Pullman, WA 99164-2930   Pullman, WA 99164-2930

Acknowledgement: The research was supported in part by the National
Science Foundation Grant No. MCS 821-7053.

International Series of
Numerical Mathematics, Vol. 77
© 1986 Birkhäuser Verlag Basel

INVERSE PROBLEMS FOR PARABOLIC PARTIAL DIFFERENTIAL EQUATIONS

P. C. DuChateau

__Abstract__. A simple strategy for dealing with "small" inverse problems is proposed. A keypoint of this strategy is the treatment of the direct problem in the weak setting, employing the strategy of judicious choosing of test functions in order to extract detailed information regarding the properties of the weak solution. These notions are illustrated by an example inverse problem.

1. Inverse Problems for Parabolic PDE's

In this discussion we propose to describe techniques suitable for treating inverse problems associated with pde's of parabolic type. The problems are "small" in the sense that they involve just a single pde and not a system, and the unknown ingredient to be identified is always a function of just a single variable. However, it is reasonable to expect that studying such simple problems may provide insight into the larger, more complicated problems which appear in practice.

Each of the problems we can consider begins with a well-posed problem for a parabolic pde:

$$(1.1) \qquad P(x,t;\partial_x,\partial_t)u(\underline{x},t) = F(\underline{x},t) \text{ in } Q_T=\{\underline{x},t: \underline{x} \in \Omega, \ 0<t<T\}$$

$$(1.2) \qquad u(\underline{x}, 0) = \Phi(\underline{x}), \qquad \underline{x} \in \Omega \subset R^N$$

$$(1.3) \qquad BC[u] = g(\underline{x}, t), \qquad \underline{x} \in \partial\Omega, \ 0 < t < T.$$

Here (1.1) represents a parabolic pde (possibly nonlinear) while (1.2), (1.3), respectively, are an initial condition and suitable boundary condition. Together, (1.1), (1.2) and (1.3) comprise a well-posed problem if a unique solution $u = u(\underline{x}, t)$ exists and depends continuously on the data. We shall refer to the functions $F(\underline{x}, t)$, $\Phi(\underline{x})$ and $g(\underline{x}, t)$ as the __proper data__ in this problem and the problem of finding $u(\underline{x}, t)$ from $\{F, \Phi, g\}$ will be referred to as the __direct problem__. When $u(\underline{x}, t)$ is known, we may define (calculate) some functional of $u(\underline{x}, t)$

$$(1.4) \qquad M[u(\underline{x}, t)] = h$$

and then, given {F, Φ, g} together with h, seek to determine some unknown ingredient such as a coefficient which appears in the equation or a boundary condition. This problem will be referred to as the inverse problem. We list now some examples of inverse problems.

Example 1. (Unknown Source Term)
$$\partial_t u(\underline{x}, t) - \nabla^2 u(\underline{x}, t) = F(u(\underline{x}, t)), \quad \underline{x} \in \Omega, \ t > 0$$
$$u(\underline{x}, 0) = \Phi(\underline{x}), \quad \underline{x} \in \Omega$$
$$u(\underline{x}, t) = g(\underline{x}, t), \quad \underline{x} \in \partial\Omega, \ t > 0$$
$$\partial_N u(\underline{x}_0, t) = h(\underline{x}_0, t) \text{ for fixed}, \quad \underline{x}_0 \in \partial\Omega, \ t > 0.$$
The proper data in this problem consists of the functions Φ and g and the overspecification is $h(\underline{x}_0, t)$ for $\underline{x}_0$ fixed. Then the inverse problem amounts to finding the unknown source term F(u) in the equation from Φ, g and h.

Example 2. (Unknown Coefficients)
$$C(u)\partial_t u = \nabla_x(K(u)\nabla_x u), \quad \underline{x} \in \Omega, \ t > 0$$
$$u(\underline{x}, 0) = \Phi(\underline{x}), \quad \underline{x} \in \Omega$$
$$K(u)\nabla_N u = g(\underline{x}, t), \quad \underline{x} \in \partial\Omega, \ t > 0$$
$$u(\underline{x}_i, t) = h_i(t), \quad i = 1, \ldots, M, \ t > 0.$$
The functions Φ and g are the proper data in this problem, while the overspecification consists of the M-functions $h_1(t), \ldots, h_M(t)$ corresponding to measurements of u at M distinct (and fixed) points on the boundary. The inverse problem consists then of finding the unknown coefficients C(u), K(u), given Φ, g and $h_1, \ldots, h_M$.

Example 3. (Unknown Boundary Coefficient)
$$\partial_t u - \nabla^2 u = 0, \quad \underline{x} \in \Omega, \ t > 0$$
$$u(\underline{x}, 0) = \Phi(\underline{x}), \quad \underline{x} \in \Omega,$$
$$\partial_N u - p(u) = g(\underline{x}, t), \quad \underline{x} \in \partial\Omega, \ t > 0$$
$$u(\underline{x}_0, t) = h(t), \quad \underline{x}_0 \in \partial\Omega \text{ (fixed)}, \ t > 0.$$
Here the proper data consists of Φ and g while h(t) is the overspecified data amounting to a measurement of u at a fixed point $\underline{x}_0$ on the boundary. The inverse problem then amounts to

finding the unknown coefficient p which appears in the boundary condition from Φ, g and h.

The strategies applied to inverse problems fall roughly into two categories: Those which compute exactly a function which approximately solves the inverse problem and those which try to compute approximately a function which exactly solve the inverse problem. The discussion here focuses on a strategy of the latter type.

2. <u>Discussion of the Approach</u>

The approach we are describing may be divided into three phases. We shall refer to the unknown ingredient always by K, the proper data (which probably consists of several functions and will therefore be a vector) by $\underline{G}$ and the overspecified data (also consisting possibly of several functions) by $\underline{H}$. The solution will be denoted by u.

<u>Phase I</u>. (The Direct Problem) We must first define:

A_0 – admissible class for K

G_0 – allowable data class

S – solution class for u.

Then we must be able to prove:

<u>Solvability Theorem</u>. For each admissible K and every allowable data set $\underline{G}$ there exists a unique u in S.

Without a theorem of this sort, we are not yet in a position to begin considering the inverse problem. Moreover, it is one of the main contentions of this discussion that the appropriate setting for this theorem is the weak (variational) formulation of the parabolic problem. Detailed information about the solution u of the direct problem is essential for a successful resolution of the inverse problem and this information is to be obtained by means of integral identities involving carefully selected test functions. These techniques are used in place of maximum principle arguments which are the tools most frequently used in connection with a classical treatment of the direct problem. In order to illustrate what is

54

intended here, consider an inverse problem for which the direct problem is the following nonlinear diffusion problem:

$$\partial_t u(x, t) = \partial_x(a(u)\partial_x u), \quad 0 < x < 1, \ 0 < t < T$$
$$u(x, 0) = u_0(x), \quad 0 < x < 1,$$
$$-a(u)\partial_x u(0, t) = g(t), \quad 0 < t < T,$$
$$\partial_x u(1, t) = 0, \quad 0 < t < T.$$

Here we suppose $a(u)$, the unknown coefficient to be determined, is a continuous positive function of a single variable, and $g \in L^2(0, T)$. Then this problem has, in general, no classical solution but it does have a weak solution $u(t) = u(\cdot, t)$ which satisfies the equation

$$(2.1) \qquad \int_0^T (u_t(t), v)_{L^2(0,1)} + a(u, v)dt = \int_0^T g(t)v(0, t)dt$$

for all v in $L^2(0T: H^1(0, 1))$. Here

$$a(u, v) = \int_0^1 a(u(x, t))\partial_x u \, \partial_x v \, dx.$$

Now suppose $\phi = \phi(x, t)$ denotes a solution of

$$\partial_t \phi + k(x, t)\partial_{xx}\phi = F(x, t), \quad 0 < x < 1, \ 0 < t < T,$$
$$(2.2) \qquad \phi(x, T) = 0$$
$$\phi(0, t) = \phi(1, t) = 0.$$

It is classical that for $k(x, t) \in C$ and $f \in L^2$, this problem has a unique solution $u \in L^2(0T: H^2(0, 1))$. Then taking $v = \phi_x \in L^2(0T: H^1(0, 1))$ in the equation (2.1) and making use of

$$(2.3) \qquad \int_0^T (u_t, \phi_x)_{L^2(0,1)} dt = \int_0^T (u_x, \phi_t)_{L^2(0,1)} dt$$

we find, from (2.1) and (2.2),

$$\int_0^T\int_0^1 \partial_x u[\partial_t \phi + k(x, t)\partial_{xx}\phi]dxdt = \int_0^T g(t)\partial_x \phi(0, t)dt$$

i.e.

$$(2.4) \qquad ((u_x, F))_{L^2(0T\times(0,1))} = (g, \partial_x\phi(0, \cdot))_{L^2(0T)}.$$

Now it is not hard to show about (2.2) that $F \geq 0$ in $(0, 1) \times (0, T)$ implies $\partial_x\phi(0, t) \leq 0$ in $(0, T)$. Using this in (2.4) leads to the result

$g \geq 0$ in $(0, T)$ implies $u_x \leq 0$ in $(0, 1) \times (0, T)$. Further refinements of this result are possible (cf. [3]) but the general observation is that rather detailed information regarding the behavior of the weak solution of the direct problem (2.1) is available by considering appropriate auxiliary problems like (2.2).

Finally, after proving a solvability result for the direct problem one must characterize the overspecified data in the problem by the following sort of lemma.

Lemma. For K in A_0 and G in D_0 it follows that u belongs to S. This in turn implies that $M[u]$ belongs to the class D_1.

D_1 = characterization of overspecified data class. Then we define: $\{G, H\}$ are consistent data for the inverse problem if $G \in D_0$ and $H \in D_1$. Note that what this lemma characterizes is the range of the mapping

$$\{A_0 \times D_0\} \ni (K, G) \quad \sim\sim\xrightarrow{\Psi} \quad H \in D_1$$

which carries the coefficient and the proper data onto the overspecification.

Phase II. (The Inverse Problem) We begin this phase be defining a solution to the inverse problem:

Defn: If $\{G, H\}$ denote consistent data for the inverse problem then the function pair, $\{K, u\}$ is a solution of the inverse problem if

    i) $K \in A_0$

    ii)   u is the unique solution of the direct problem for proper data G

    iii) $M[u] = H$.

Next we must prove:

Theorem (**Sufficiency of the Data**) Let $\{G, H\}$ denote consistent data for the inverse problem and suppose $\{K_1, u_1\}$, $\{K_2, u_2\}$ each is a solution of the inverse problem having data $\{G, H\}$. Then $K_1 = K_2$ in the natural topology of A_0.

The implication of this theorem is that the data $\underline{H}$ is sufficient for the determination of K if distinct K's cannot produce the same response $\underline{H}$ to identical input $\underline{G}$.

Phase III. (Approximation of K from $\underline{G}$, $\underline{H}$) A second contention of this discussion is that we have postulated some model leading to a well-posed problem for a parabolic pde and that there **exists** some K in A_0 which produces the output $\underline{H}$ from input $\underline{G}$. Therefore, no **existence proof** for K is needed. Instead, the more natural question to consider is whether K can be **approximated** for $\underline{G}$ and $\underline{H}$ and what sort of estimates are obtainable.

We will now illustrate these general remarks by means of the following example.

3. Example Inverse Problem

Consider the following direct problem

$$U(y)\partial_x T(x, y) = \partial_{yy} T(x, y), \quad \begin{matrix} 0 < x < L \\ 0 < y < h \end{matrix}$$

(3.1) $\quad T(0, y) = P(y), \quad 0 < y < h,$

$$T(x, 0) = T_C, \; T(x, h) = T_H > T_C, \quad 0 < x < L,$$

together with the overspecification

(3.2) $\quad T(L, y) = Q(y), \; 0 < y < h,$

and suppose we are to determine the unknown coefficient $U(y)$, $0 \leq y \leq h$. For a complete discussion of this problem together with its physical motivation, see [1] and [2].

First, define

$$A_0 = \{U \in H_0^1(0, 1): U(y) > 0, \quad 0 < y < h\}$$

$$D_0 = \{P \in H^1(0, 1): P(0) = T_C, \; P(h) = T_H > T_C\}.$$

Then for $\ell(y) = T_C + (T_H + T_C)y/h$, $0 \leq y \leq h$, let

(3.3) $\quad Z(x, y) = T(x, y) - \ell(y).$

If $T(x, y)$ satisfies (3.1) then $Z(x, y)$ satisfies:

$$(3.4) \quad (\partial_t Z, \phi)_W + \int_0^1 \partial_x Z \, \partial_x \phi \; dx = 0 \quad \forall \phi \in L^2(0T, H^1), \; Z(0) = P - \ell$$

where

$$(\phi, \psi)_W = \int_0^1 U(y)\phi\psi \, dy.$$

Then it is classical that

For each U in A_0 and every P in D_0 there is a unique Z in $L^2(OT: H_0^1(0, h))$ satisfying (3.4). Then $T = Z + \ell$ will be defined as the unique solution of the direct problem.

It is then clear that

$$D_1 = \{Q \in H^1(0, h): Q(0) = T_C, \ Q(h) = T_H\}.$$

Finally, $\{U, T\}$ is a solution of this inverse problem if U lies in A_0, T is a solution of the direct problem and $T(L, y) = Q(y)$, $0 \leq y \leq h$.

Then we can prove:

Theorem. (Sufficiency of the Data) If $\{P, Q\}$ **are consistent data for this inverse problem and if** $\{U, T\}$, $\{V, S\}$ **denote two solutions of the inverse problem then** $U(y) = V(y)$ **in a neighborhood of each point** $y_0 \in (0, h)$ **where** $T(x, y_0)$ **is a strictly monotone function of** x.

The main idea of the proof of this result is as follows. If $\{U, T\}$, $\{V, S\}$ each solve this inverse problem, then for arbitrary $\phi = \phi(x, y)$ in $L^2(OT: H_0^1)$,

$$(3.5) \qquad \int_0^L \int_0^h (T - S)[v\partial_x\phi + \partial_{yy}\phi] = \int_0^h \int_0^L (u - v)\phi\partial_x T.$$

Now if $y_0 \in (0, h)$ is such that $T(x, y_0)$ is monotone in x, then we choose the test function ϕ such that

$$[v\partial_x\phi + \partial_{yy}\phi] = 0 \qquad (0 < x < L, \ 0 < y < h),$$
$$\phi(x, 0) = \phi(x, h) = 0.$$

In addition, by appropriate choice of $\phi(L, y)$, we can ensure that the product $\phi\partial_x T$ is nonnegative in a neighborhood of y_0 and is zero elsewhere. Then (3.5) implies $U(y) = V(y)$ in a neighborhood of y_0.

Evidently, if the temperature field $T(x, y)$ is such that T is a monotone function of x for each y in (0, h) then the temperature profiles carry sufficient information to determine

U(y), the velocity distribution. In order to approximate U(y) from P and Q note that for $\phi = \phi(y) \in H_0^1(0, h)$,

$$\int_0^L \int_0^h \phi U \partial_x T = \int_0^h \phi U[Q - P]dy$$

and

$$\int_0^L \int_0^h \phi \partial_{yy} T = \int_0^L \int_0^h \phi''(y)T \sim \frac{L}{2}\int_0^h \phi''[Q + P - 2\ell(y)]dy.$$

That is,

(3.6) $$\int_0^h \phi U[Q - P]dy \sim \frac{L}{2}\int_0^h \phi''[Q + P - 2\ell(y)]dy.$$

Then if $\{\phi_K(y)\}$ denotes a basis in $H_0^1(0, h)$, we can write

$$U[Q - P] = \sum_k (U[Q - P], \phi_K)_{L^2(0,h)} \phi_K(y).$$

But according to (3.6) the $L^2(0, h)$ projections of $U[Q - P]$ on each of ϕ_K can be approximated by integrals involving only the data, P, Q and $\ell(y)$. Moreover, the observation that

$$(Q - P)(y) = \partial_x(T(\hat{x}(y), y)L$$

leads to estimates for the difference $U - U^*$ between the exact distribution U(y) and its approximant. These estimates are of the form

(3.7) $$\|U - U^*\| \leq \frac{C_1}{L} + D_2 L^2.$$

Recalling that L denotes the spacing between the positions where the data P and Q are collected, we see that (3.7) implies the existence of an "optimal" choice for this spacing. If L is too small, the approximation error is dominated by data noise while if L is too large the integral approximation used in (3.6) causes the accuracy of the algorithm to deteriorate. As in the proof of the sufficiency of the data, we see that monotonicity of the temperature as a function of x plays a significant role in the approximation of U(y).

A complete treatment of this example problem can be found in [1].

4. Summary

We are proposing here that "sufficiently small" inverse problems can be treated according to the following program:

a) **The Direct Problem**

Careful treatment of the direct problem allows characterization of the range of the coefficient map:

$$A_0 \times D_0 \ni (K, \underline{G}) \;\; \sim\!\sim\!\xrightarrow{\;\Psi\;}\; \underline{H} \in D_1 .$$

b) **The Inverse Problem**

"Sufficiency of the data" amounts to proving that the "section" $\Psi(\cdot, \underline{G})$ from A_0 to D_1 is injective.

c) **The Approximation**

Assuming $\underline{H}$ lies in the range of Ψ, then an approximate inverse to $\Psi(\cdot, \underline{G})$ can be constructed,

$$A_0 \ni K \xrightarrow{\;\Psi\;} \Psi(K, \underline{G}) \subset \underline{H} \in D_1$$

$$A_N \ni K^* \xleftarrow{} \Psi^{-1}_{G,N}(\cdot) = \text{approximate inverse.}$$

Finally, we propose that the direct problem be considered in the weak setting and that detailed information regarding the properties of the weak solution be extracted by the strategy of choosing appropriate test functions.

REFERENCES

1. DuChateau, P., Moaveni, S. and Tan, L., An inverse problem for indirect determination of a velocity distribution, preprint.

2. DuChateau, P., Moaveni, S. and Burns, P., Indirect measurement of velocity distributions, in preparation.

3. DuChateau, P., An inverse problem for a nonlinear parabolic PDE, preprint.

International Series of
Numerical Mathematics, Vol. 77
© 1986 Birkhäuser Verlag Basel

THREE-DIMENSIONAL INVERSE SCATTERING

Margaret Cheney, James H. Rose and Brian DeFacio

Abstract. We consider the problem of obtaining information about an inaccessible region of space from scattering experiments. Inverse scattering theory for the time-independent Schrödinger equation $[\Delta + h^2 - V(x)]\Psi(k,x) = 0$ is summarized. It is most easily understood by considering the associated hyperbolic equation $[\Delta - \partial_{tt} - V(x)]u(t,x) = 0$. Particular attention is paid to those aspects of the theory that hold for the wave equation $[\Delta - n^2(x)\partial_{tt}]u(t,x) = 0$.

The inverse scattering problem is concerned with obtaining information about an inaccessible region of space by means of scattering experiments. One passes some kind of wave through the inaccessible region, and measures the scattered wave on a sphere (with radius finite or infinite) surrounding the region.

This paper will outline the three-dimensional inverse scattering theory for the time-independent Schrödinger equation

$$(1) \qquad [\Delta + k^2 - V(x)]\Psi = 0 \ .$$

here the potential $V(x)$ describes the inaccessible region of space, and k^2 is the energy of our probing wave. Inverse scattering theory for (1) is most easily understood [1] by also considering the related hyperbolic equation

$$(2) \qquad [\Delta - \partial_{tt} - V(x)]u = 0 \ .$$

Two inverse scattering methods will be considered, and the second of these will apply also to the reduced wave equation

$$(3) \qquad [\Delta + k^2 n^2(x)]\Psi = 0$$

and its Fourier transform

$$(4) \qquad [\Delta - n^2(x)\partial_{tt}]u = 0 \ .$$

If we write $n^2(x) = 1 - V(x)$, we see that (1) and (3) differ only by a factor of k^2 in front of $V(x)$. For all these problems, we will assume that $V(x) \to 0$ sufficiently rapidly at infinity.

We begin by considering equations (1) and (2). They are clearly related by the Fourier transform

$$(5) \qquad u(t,x) = \frac{1}{2\pi} \int_{-\infty}^{\infty} e^{-ikt} \, \psi(k,x) dk \; .$$

For simplicity, we assume that (1) supports no bound states, so that no difficulties arise with (5).

Considering equation (1), we probe our unknown region with a plane wave travelling in direction $e \in S^2$: $\psi_{inc} = e^{ike \cdot x}$. We define solutions of (1) by means of the Lippmann-Schwinger integral equation [2]

$$(6\pm) \qquad \psi^{\pm}(K,ex) = e^{ike \cdot x} + \int G_0^{\pm}(k,(x-y))V(y)\psi^{\pm}(k,e,y)dy,$$

where

$$(7) \qquad G_0^{\pm}(k,r) = -(4\pi r)^{-1}\exp(\pm ikr)$$

are Green's functions for the operator $\Delta + k^2$. G^+ corresponds to an outgoing radiation condition at infinity, whereas G^- corresponds to an incoming one. The two solutions ψ^+ and ψ^- are related by

$$(8) \qquad \psi^+(k,e,x) = \psi^-(-k,-e,x) \; .$$

Equation (6) allows us to solve the direct scattering problem for (1).

The direct scattering problem for (2) can be solved in a similar way. We take the incident wave to be $u_{inc} = \delta(t-e \cdot x)$, and we solve (2) by means of the equation

$$(9\pm) \quad u^{\pm}(t,e,x) = \delta(t-e \cdot x) + \iint_{-\infty}^{\infty} \hat{G}_0^{\pm}(t-\tau,|x-y|)V(y)u^{\pm}(\tau,e,y)d\tau \, dy$$

where

$$(10) \qquad \hat{G}_0^{\pm}(t,r) = -(4\pi r)^{-1}\delta(r \mp t)$$

is the Fourier transform of (7). Again u^+ and u^- are related by

$$(11) \qquad u^+(t,e,x) = u^-(-t,-e,x) \; .$$

We now turn our attention to the inverse scattering problem for (1) and (7). For this it is useful to use the large-k asymptotic expansion [3] of (6+)

(12) $\psi^+(k,e,x) = \exp(ike \cdot x)[1 + B(e,x)(ik)^{-1} + \ldots]$.

For (2), this corresponds to the progressing wave expansion [4]

(13) $u^+(t,e,x) = \delta(t-e \cdot x) + B(e,x)H(t-e \cdot x) + $ (smoother terms),

where H denotes the Heaviside function that is one for positive arguments and zero for negative arguments.

To determine the coefficient B, we substitute (13) in (2) and equate coefficients of corresponding singularities. Equating the coefficients of δ'' and δ' result in trivial equations. Equating the coefficients of δ, however, results in [5]

(14) $$2e \cdot \nabla_x B(e,x) = -V(x) \ .$$

Equation (14) is very useful. We note that the left side is a directional derivative. If we integrate (14) along a line with direction e, we therefore have

(15) $$2[B(e,P_1) - B(e,P_2)] = -\int_{\text{line } P_1 P_2} V(x)dx$$

where P_1 and P_2 are as in Figure 1.

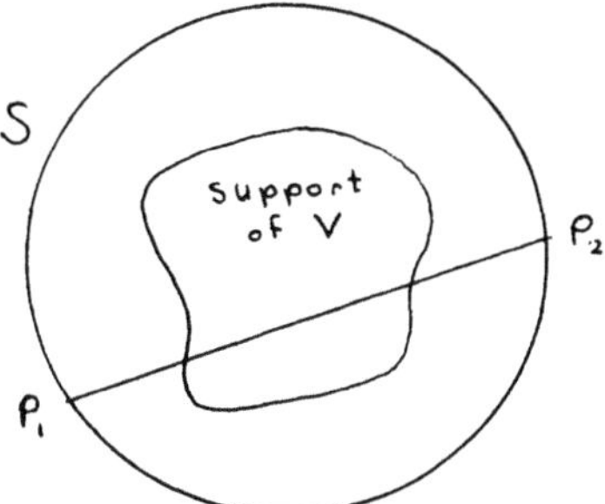

Figure 1.

The quantity B can be easily obtained as follows. We probe the unknown region with the plane wave $\delta(t - e \cdot x)$. On all points of a large sphere S, we measure the wave field u for a long time interval. At each point P on the sphere, the wave field u behaves as in Figure 2. At first there is no signal, because the disturbance has not yet reached the point P. Then at time $t = e \cdot P$, the δ function passes through, and u behaves according to (13). The quantity B(e,P) is just the height of the jump in u at t = e · P.

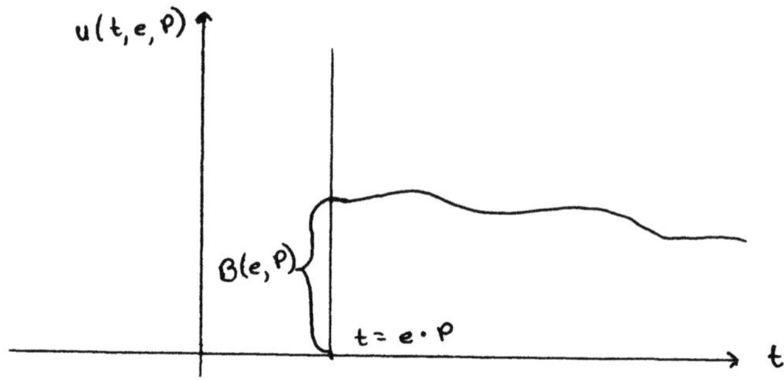

Figure 2.

Equation (15) thus gives us the following inversion
algorithm [6,7,1]. For each incident wave, we measure B at
all points on the sphere S. We thus obtain B(e,P) for all e
and P ∈ S^2. According to (15), this gives us all line integrals
of V(x). The function V can easily be recovered from its
line integrals by means of the Radon transform inversion formula. Indeed, this problem is solved every day in hospitals
around the world as part of the CAT-scan procedure.

This inversion method is closely related to the method
based on the Born approximation. It therefore has the same dis-
advantage as the Born inversion method; namely, it relies on
accurate high-frequency data, which in many cases is not available. Also, it relies heavily on the fact that the characteristics of (2) are straight lines. This method thus does not apply
to (4). We therefore look for another inversion method.

The above method relied on the large-k expansion (12)
for (6+); we now consider the large-|x| expansion of (9+). It
is

$$(16) \qquad u^+(t,e,x) = \delta(t-e\cdot x) + \frac{R(t-|x|,\hat{x},e)}{|x|} + \dots ,$$

where $\hat{x} = x/|x|$ and R is the <u>impulse response</u>

$$(17) \qquad R(t,e,e') = -\frac{1}{8\pi^2} \int V(y)u^+(t+e\cdot y,e',y)dy .$$

R is the Fourier transform of the usual scattering amplitude for the Schrödinger equation (1).

We will consider the inverse problem of obtaining V
from R. We note that this is an overdetermined problem, because R depends on five variables whereas V depends on only
three. We therefore also have the problem of characterizing
those functions of five variables that can arise as impulse re-
sponses. This characterization problem is currently unsolved in
the sense that the only way to tell whether R is admissible is
to go through the entire inversion procedure below. If, at the
end, it doesn't work, then R was inadmissible.

Our plan for the inversion procedure is as follows.
First we will find an integral equation [8] relating R and u.
We solve this equation for u, and then from u we obtain B
via (13), and from B we obtain V via (14).

In obtaining [9,10] an integral equation relating R
and u, we will interpret $\hat{G}_0^+$ as propagating waves forward in
time, and $\hat{G}_0^-$ as propagating waves backwards in time. We can
thus interpret (9+) as saying that u^+ is the solution of (2)
that looked like $\delta(t - e \cdot x)$ "in the past", and (9-) as saying
that u^- is the solution of (2) that will look like $\delta(t - e \cdot x)$
"in the future".

Next we do a very short computation using (17) and
(9+):

$$\int \dot{R}(t-e\cdot x,e,e')de = -\frac{1}{4\pi} \int V(y) \int \dot{u}(t-e\cdot(x-y),e',y)de \, dy$$

$$= -2\pi \int V(\hat{G}_0^+ - \hat{G}_0^-)u^+ \, d\tau \, dy$$

$$= -2\pi[u^+ - \delta - \iint \hat{G}_0^- V u^+ \, dy \, d\tau],$$

where $\cdot$ denotes d/dt. In the second step we have merely done
the e integration and written the result in terms of $\hat{G}_0^+$ and
$\hat{G}_0^-$. We rewrite this result in a more suggestive way:

$$(18) \quad u^+(t,e,x) = \delta(t-e\cdot x) - \frac{1}{2\pi} \int \dot{R}(t-e'\cdot x,e',e)de' +$$

$$\iint_{-\infty}^{\infty} \hat{G}_0^-(t-\tau,|x-y|)V(y)u^+(\tau,e,y)d\tau \, dy \, .$$

We interpret (18) as telling us that u^+ is the solution of (2) that looks like

(19) $\qquad \delta(t-e\cdot x) - \frac{1}{2\pi} \iint \dot{R}(t-\tau,e',e)\delta(\tau-e'\cdot x)de'\ d\tau$

"in the future". But we know that the solution that looks like $\delta(t-e\cdot x)$ "in the future" is u^-. By the superposition principle, (18) tells us that

(20) $\qquad u^+(t,e,x) = u^-(t,e,x)$
$$- \frac{1}{2\pi} \int_{S^2} \int_{-\infty}^{\infty} \dot{R}(t-\tau,e',e)u^-(\tau,e',x)d\tau\ de'\ .$$

Equation (20) is an integral equation relating u and R. It can be converted into a more useful integral equation by the following steps [1]. First, we write

(21) $\qquad u^{\pm}(t,e,x) = \delta(t-e\cdot x) + u_{sc}^{\pm}(t,e,x)\ .$

Next we note that u^+ is causal:

(22) $\qquad u^+(t,e,x) = 0 \quad \text{for} \quad t < e\cdot x\ .$

We use (21), (22), and (11) to convert (20) into the following Fredholm II integral equation on the region $t > e\cdot x$:

(23) $\qquad u_{sc}^+(t,e,x) = -\frac{1}{2\pi} \int_{S^2} \dot{R}(t - e'\cdot x,e',e)de'\ +$
$$- \frac{1}{2\pi} \int_{S^2} \int_{e'\cdot x}^{\infty} \dot{R}(t+\tau,-e',e)u_{sc}^+(\tau,e',x)d\tau\ de'\ .$$

Newton has shown [11] that under very mild conditions this Fredholm II equation can always be solved (even if R contains bad data). Once the solution u^+ is known, V can be recovered from (13) and (14).

If R is admissible, then the left side of (14), which is computed from the solution of (20), will be independent of e. This is called the "miracle" [8]. It characterizes admissible R's.

This completes the outline of three-dimensional inverse scattering theory for (2). Next we consider the modification necessary to treat (4). In most equations above, we need merely insert ∂_{tt} after the V. We will denote the equations so obtained by ($\ddot{\ }$). Thus we consider scattering solutions defined by ($\ddot{9}\ddot{\pm}$). R is defined by ($\ddot{17}$). We derive ($\ddot{18}$) exactly as before, and from it we derive (20). Equation (20) has exactly the same form for (4) as it does for (2) [12].

The derivation of (23) from (20), however, is different for (4). Equation (22) is only true if waves propagate with speed one or slower, i.e., if $1 - V(x) \geq 1$ everywhere. In other words, the supports of u^+ and u^- are disjoint only if $V(x) \leq 0$ everywhere.

We can again define u_{sc} by (21) and (if $V \leq 0$) obtain (23), but we should keep in mind that u_{sc} also contains delta functions. Thus (23) is no longer a Fredholm II equation on L^2, and little is known about it.

If by some means (23) can be solved for u^+, then V can be recovered, but not by (14). The expansion corresponding to (13) in this case is

$$(24) \quad u^+(t,e,x) = z_0(e,x)\delta(t-s(e,x)) + z_1(e,x)H(t-s(e,x))$$
$$+ \text{(smoother terms)}$$

where s satisfies the eikonal equation

$$(25) \qquad |\nabla s(e,x)|^2 = n^2(x)(= 1 - V(x))$$

and z_0 and z_1 satisfy the usual transport equations of geometrical optics. Thus, if u^+ is known, then the speed ∇s of the delta function is also known, and V can be recovered from (25).

REFERENCES

1. J. H. Rose, M. Cheney, and B. DeFacio, The connection between time- and frequency-domain three-dimensional inverse scattering methods, J. Math. Phys. **25** (1984), 2995-3000.

2. R. G. Newton, <u>Scattering Theory for Waves and Particles,</u> 2nd edition, Springer, New York (1982).

3. M. Cheney, A rigorous derivation of the 'miracle' identity of three-dimensional inverse scattering, J. Math. Phys. **25** (1984), 2988-2990.

4. R. Courant and D. Hilbert, <u>Methods of Mathematical Physics,</u> Volume II, Wiley, New York (1962).

5. C. S. Morawetz, A formulation for higher dimensional inverse problems for the wave equation, Comp. & Maths. with Appls. **7** (1981), 319-331.

6.   C. J. Callias and G. A. Uhlmann, Asymptotics of the scatter-
      ing amplitude for singular potentials, technical report
      MSRI 034-83, Mathematical Sciences Research Institute,
      Berkeley.

7.   B. DeFacio and J. H. Rose, Exact inverse scattering theory
      for the non-spherically symmetric three-dimensional plasma
      wave equation, Phys. Rev. A, 31 (1985), 897-902.

8.   R. G. Newton, Inverse scattering II. Three dimensions, J.
      Math. Phys. 21 (1980), 1698-1715; 22 (1981), 631; 23
      (1982), 693.

9.   J. H. Rose, M. Cheney, and B. DeFacio, Determination of the
      wavefield from the scattering data, Phys. Rev. Lett., in
      press.

10.   M. Cheney, J. H. Rose, and B. DeFacio, On the direct rela-
      tion of the wavefield to the scattering amplitude, in
      preparation.

11.   R. G. Newton, Variational principles for inverse scatter-
      ing, Inverse Problems 1 (1985), 371-380.

12.   J. H. Rose, M. Cheney and B. DeFacio, Three-dimensional in-
      verse scattering: plasma and variable velocity wave equa-
      tions, J. Math. Phys. 26 (1985), 2803-2813.

Margaret Cheney   James H. Rose
Department of Mathematics   Center for Nondestructive Evaluation
Duke University   Iowa State University
Durham, NC 27706   Ames, Iowa 50011
USA   USA

Brian DeFacio
Department of Physics and Astronomy
University of Missouri
Columbia, MO 65211

Acknowledgement. The work of MC was partially supported by ONR
contract number N00014-85-K-0224. JHR's work was supported by
the NSF University/Industry Center for NDE at Iowa State University.

International Series of
Numerical Mathematics, Vol. 77
© 1986 Birkhäuser Verlag Basel

ON ISOSPECTRAL GRADIENT FLOWS - SOLVING
MATRIX EIGENPROBLEMS USING DIFFERENTIAL EQUATIONS

Kenneth R. Driessel

Abstract In this paper I shall consider the following problem: Given a symmetric (or Hermitian) matrix, find its eigenvalues. I shall use the theory of ordinary differential equations to solve this problem. In particular, I shall describe a spectrum-preserving dynamical system on symmetric (or Hermitian) matrices that flows "downhill" toward diagonal matrices. I shall show that diagonal matrices are the only stable equilibrium points of this flow.

Contents

Introduction

An Isospectral Gradient Flow for Real Symmetric Matrices

An Isospectral Gradient Flow for Complex Hermitian Matrices

Concluding Remarks

References

Acknowledgements

Introduction

In this paper I shall consider the following "spectral" problem: Given a symmetric (or Hermitian) matrix, find its eigenvalues. I shall describe a spectrum-preserving ("isospectral") ordinary differential equation on symmetric (or Hermitian) matrices that provides a solution to this problem. I shall show that diagonal matrices are the only stable equilibrium points of this dynamical system. Consequently, the integral curves of this differential equation generally flow from an initial matrix toward its diagonal matrix of eigenvalues. In this way I obtain a highly constructive proof of the spectral theorem. This ordinary differential equation can be used as the basis for a numerical algorithm.

In a paper [4] which appeared in 1983, Dieft, Nanda and Tomei suggested that numerical eigenvalue algorithms might be based on the dynamical system known as the Toda flow. (See also the paper [10] by Nanda which also appeared in 1983. For earlier work on the Toda flow see, for example, the 1974 paper [5] by Flascka, the 1975 paper [9] by Moser, and the references in the other papers cited in this paragraph.) Earlier Symes had recognized the connection between the Toda flow and the QR-algorithm. (See papers [13] and [14] which appeared in 1980 and 1982 respectively.) In 1984 expository papers by Watkins [16] and Chu [1] appeared; these two papers further explain this connection. In fact the QR-algorithm in its simplest form is a discrete version of the continuous Toda flow. The Jacobi algorithm is a discrete version of the continuous gradient system that I describe in this paper. (See, for example, the 1980 book [11] by Parlett and/or the 1983 book [6] by Golub and Van Loan for descriptions of the Jacobi algorithm.)

The paper [4] by Dieft, Nanda and Tomei includes a study of the equilibrium points of the Toda flow. (See Theorems 3 and 4 and the remarks preceding Theorem 4 in their paper.) They show that all of the equilibrium points are diagonal matrices Diag $(d_1, d_2, \ldots, d_n)$ where $(d_1, d_2, \ldots, d_n)$ is permutation of the eigenvalues $\lambda_1 > \lambda_2 > \ldots > \lambda_n$ associated with the given isospectral surface. However, they show further that the diagonal matrix Diag $(\lambda_1, \lambda_2, \ldots, \lambda_n)$ with the eigenvalues in descending order on the diagonal is the only stable equilibrium point. It follows that the Toda system flows away from the other diagonal matrices. I was motivated by this algorithmically undesirable property to design the gradient flow that I present in this paper. For this gradient system all the diagonal matrices of eigenvalues are stable equilibrium points (and there are no others).

Remark (On using ordinary differential equations for algorithm design) In remarks like this one, I discuss tangential topics. The hurried reader may skip them.

Any algorithm that finds eigenvalues, necessarily finds the roots of (characteristic) polynomials. Such an algorithm must be iterative. Consequently, it describes a dynamical system. Therefore, we expect that the theory of ordinary differential equations (which is the best developed branch of dynamical systems theory) should be useful when we design such an algorithm. Since continuous problems are generally easier than discrete problems, we should consider differential equations before considering related discrete dynamical systems.//

<u>Remark</u> (On motivation and connection with inverse problems)

Let X be a fixed positive real number and let a : $[0,X] \times [0,X] \to R$ be a smooth symmetric kernel (i.e. a real-valued function $(x,y) \to a(x,y)$ defined on the square $[0,X] \times [0,X] \subseteq R \times R$ satisfying $a(x,y) = a(y,x)$.). Let $L_2[0,X]$ be the Hilbert space of square-integrable real-valued functions defined on the interval $[0,X] \subseteq R$ with inner product

defined by $<f,g> := \int_0^X dx\ f(x)g(x)$. The kernel $a(x,y)$ determines an integral operator:

$$A : L_2[0,X] \to L_2[0,X] \quad \text{defined by}$$

$$Af(x) := \int_0^X dy\ a(x,y)f(y).$$

We are interested in the following "inverse" problem ("Fredholm integral equation of the first kind"): Given function h, find function f such that Af = h. (One may discretize this problem to obtain an equation Af = h where A is a symmetric N-by-N matrix and f and h are vectors.) This problem is generally "ill-posed". (The corresponding matrix problem is generally "ill-conditioned".) In particular, there generally exists ("spectral decomposition") a sequence of "eigenvalue-function" pairs (λ_m, f_m) satisfying

$$Af_m = \lambda_m f_m, \quad <f_m, f_n> = \delta_{mn},$$

$$|\lambda_1| \geq |\lambda_2| \geq \cdots \geq |\lambda_m| \geq \cdots,$$

$$\lim_{m \to \infty} \lambda_m = 0.$$

(In the matrix problem we have eigenvalue-vector pairs (λ_m, f_m), for m = 1,2,...,N, satisfying

$$Af_m = \lambda_m f_m, \quad <f_m, f_n> = \delta_{mn},$$

$$|\lambda_1| \geq |\lambda_2| \geq \cdots \geq |\lambda_N|, \text{ and}$$

$$|\lambda_1|/|\lambda_N| \text{ large.})$$

We may stabilize this problem by restricting our attention to the subspace span $\{f_1, f_2, \ldots, f_M\}$ where λ_1/λ_M is not too large. In order to do so we need to be able to quickly solve eigenproblems. In particular, we want to efficiently solve eigenproblems when the kernel $a(x,y)$ has additional structure (e.g. when the kernel is Toeplitz: $a(x,y) = b(x-y)$ for some function b). I hope that isospectral flow techniques will lead to such efficient algorithms.//

An Isospectral Gradient Flow for Real Symmetric Matrices

In this section, I shall consider the following ("spectral") problem: Given a real symmetric matrix, find its eigenvalues. I shall use the theory of ordinary differential equations to solve this problem. In particular, I shall describe a spectrum-preserving ("isospectral") dynamical system on symmetric matrices that flows "downhill" toward diagonal matrices. The choice of the system is based on the following idea: Look for a miminum of a real-valued function that measures the size of the off-diagonal part of a matrix. I shall show that every relative minimum occurs at a diagonal matrix.

Let R denote the set of real numbers and let $R^{n \times n}$ denote the set of all n-by-n matrices over R. Recall that the ("Frobenius") inner product $<, > : R^{n \times n} \times R^{n \times n} \to R$ is defined by $<X,Y> := \text{trace}(XY^T)$. (I use a superscript T to denote the transpose operation and I use the word "trace" to denote the trace function: $(X^T)_{ij} := x_{ji}$, trace $X := \sum x_{ii}$. Note $<X,Y> = \sum x_{ij}y_{ij}$.) We define the function

$$\text{off-diag: } R^{n \times n} \to R \text{ by } X \to (1/2) <X - \text{Diag}.X, X - \text{Diag}.X>,$$

where $\text{Diag}.X := \text{Diag}(x_{11}, x_{22}, \ldots, x_{nn})$ is the diagonal matrix with elements $x_{11}, x_{22}, \ldots, x_{nn}$ on the diagonal. (I sometimes indicate function evaluation by f.x in place of f(x).) We now compute the (Frechet) derivative of the off-diag function.

<u>Proposition</u> (Derivative of off-diag) The derivative of the off-diag function at matrix X is the linear function

$$D.off\text{-}diag.X : R^{n \times n} \rightarrow R$$

given by

$$H \rightarrow \langle \nabla.off\text{-}diag.X, H \rangle$$

where

$$\nabla.off\text{-}diag.X := X - Diag.X.$$

We shall call $\nabla.off\text{-}diag.X$ the "spatial gradient of the off-diag function at X ".

<u>Proof</u> (See, for example, Chapter 5 "Differential Calculus" in the 1969 book [8] by S. Lang, for a description of the Frechet derivative and its properties.) Using notation inspired by the computer language Lisp, we simply calculate (using the product rule for the main step):

$$D.(Y \rightarrow 1/2 \langle Y - Diag.\,Y, Y - Diag.\,Y \rangle).X.H$$

$$= 1/2 \langle D.(Y \rightarrow Y - Diag.Y).X.H, (Y \rightarrow Y - Diag.Y).X \rangle$$

$$+ 1/2 \langle (Y \rightarrow Y - Diag.Y).X, D.(Y \rightarrow Y - Diag.Y).X.H \rangle$$

$$= 1/2 \langle H - Diag.H, X - Diag.X) + 1/2 \langle X - Diag.X, H - Diag.H \rangle$$

$$= \langle X - Diag.X, H - Diag.\,H \rangle$$

$$= \langle X - Diag.X, H \rangle. \; //$$

We define the "isospectral" surface determined by a matrix X as follows:

$$Iso.X := \{U^T X U : U \in R^{n \times n} \text{ and } U^T U = I\}.$$

<u>Remark</u> Let $0(n) := \{U \in R^{n \times n} : U^T U = I\}$ be the orthogonal group and let Sym: $= \{X \in R^{n \times n} : X^T = X\}$ be the set of symmetric matrices. Define the group action $* : 0(n) \times Sym \rightarrow Sym$ by $U * X := U^T X U$. Then Iso.X is the orbit of X. //

74

We want the flow to preserve the spectrum - to stay in the isospectral surface. It should be tangent to the isospectral surface. We now compute the tangent space and the normal space.

Proposition Let X be an n-by-n matrix. The tangent space at X to the surface Iso.X is $\text{Tan.Iso.X} := \{[X,K]: K \in R^{n \times n}$ and $K + K^T = 0\}$ where $[X,K] := XK - KX$.

Remark If $K + K^T = 0$ we say that K is "skew" or "skew-symmetric". Note that if X is symmetric then the matrix

$$[X,K] = XK - KX = XK + (XK)^T$$

is also symmetric. //

Proof Consider a smooth curve $U : R \to R^{n \times n}$ satisfying $U(t) \, U(t)^T = I$ and $U(0) = I$. (Note that the image of U is contained in the orthogonal group $O(n)$.) Then

$$(d/dt)(U(t)^T X U(t))|_{t=0} = U'(t)^T X U(t) + U(t)^T X U'(t)|_{t=0}$$
$$= XK - KX$$

where $K := U'(0)$. We also have

$$0 = (d/dt) \, I = (d/dt) \, (U(t)U(t)^T)|_{t=0}$$
$$= U'(t) \, U(t)^T + U(t) \, U'(t)^T|_{t=0}$$
$$= K + K^T.$$

Thus the tangent space is contained in the set Tan.Iso.X defined in the proposition.

Now let K be any skew matrix. Define $U: R \to R^{n \times n}$ by $U(t) := \exp(Kt)$. Note that $U(t) \, U(t)^T = \exp(K + K^T)t = I$, $U(0) = I$ and $U'(0) = K$. //

Remark The following proposition shows that generally the linear map Skew $\to$ Tan.Iso.X defined by $K \to [X,K]$ is one-to-one.

Proposition Let X be a symmetric matrix with distinct eigenvalues. If K is skew and $[X,K] = 0$ then $K = 0$.

Recall (or easily check) the following result:

Lemma Let D be a diagonal matrix with distinct diagonal elements and let Y be any n-by-n matrix. If $DY = YD$ then Y is diagonal.

Proof (of proposition). Let $X = UDU^T$ where U is orthogonal and D is diagonal. We have the following sequence of implications:

$$XK = KX \Rightarrow UDU^TK = KUDU^T \Rightarrow DU^TKU = U^TKUD \Rightarrow$$

$$E := U^TKU \text{ is diagonal} \Rightarrow K = UEU^T \text{ is both symmetric and skew} \Rightarrow K = 0. \; //$$

For $X \in R^{n \times n}$, we set

$$\text{Nor.Iso.}X := \{Y \in R^{n \times n}: [X^T, Y] \text{ is symmetric}\}.$$

We call this set the "normal space" at X to the surface Iso.X. The next proposition justifies this terminology. It shows that Nor.Iso.X is the orthogonal complement of Tan.Iso.X.

Proposition Let X be an n-by-n matrix. Then

$$\text{Nor.Iso.}X = (\text{Tan.Iso.}X)^\perp.$$

We prove two lemmas which we will use in the proof of the proposition.

Lemma (Adjoint of $X \to [A,X]$) Let A, X and Y be n-by-n matrices. Then

$$\langle [A,X], Y \rangle = \langle X, [A^T,Y] \rangle.$$

Proof We calculate:

$$\text{trace}((AX - XA)Y^T) = \text{trace}(XY^TA - XAY^T)$$

$$= \text{trace}(X(A^TY - YA^T)^T)$$

The next lemma shows that the orthogonal complement in $R^{n \times n}$ of the set of skew matrices is the set of symmetric matrices (in symbols $\text{Skew}^\perp = \text{Sym}$).

Lemma Let Y be an n-by-n matrix. Then Y is symmetric if and only if Y satisfies the following condition: for all skew K, $\langle K, Y \rangle = 0$.

Proof Let K^{ij} be the matrix with 1 in position (i,j), -1 in position (j,i) and 0 elsewhere. Then $\langle K^{ij}, Y \rangle = y_{ij} - y_{ji}. \; //$

<u>Proof</u> (of proposition) We have the following equivalent conditions:

$$Y \in (\text{Tan. Iso. } X)^{\perp},$$

$$\text{for all skew } K, \; 0 = <[X,K], \; Y>,$$

$$\text{for all skew } K, \; 0 = <K, \; [X^T,Y]>,$$

$$[X^T,Y] \text{ is symmetric.} \quad //$$

<u>Corollary</u>. Let X and Y be symmetric matrices. Then $Y \in$ Nor.Iso.X if and only if XY = YX.

<u>Proof</u>. Note $[X^T, Y] = [X,Y]$ is both skew and symmetric. //

We define the "(isospectral) surface gradient" of the function off-diag at X as the projection onto the tangent space of the spatial gradient:

$$\text{Grad.off-diag.X} := \text{Projection.}(\text{Tan.Iso.X}).(\nabla.\text{off-diag.X}).$$

The function Iso.A $\to$ $R^{n \times n}$ given by X $\to$ Grad.off-diag.X defines a "gradient" vector field on Iso.A. We now consider the associated ordinary differential equation.

<u>Proposition</u> (Isospectral Gradient Flow) Let A be a symmetric matrix. Then there exists a unique smooth function X: R $\to$ Iso.A satisfying

$$X' = - \text{Grad.off-diag.X},$$

$$X(0) = A.$$

The function off-diag is nonincreasing along X. If M is an isolated minimum on Iso.A of the function off-diag then M is an asymptotically stable equilibrium point of this gradient system.

<u>Remark</u> The integral curves of the differential equation are called "gradient lines" on Iso.A of the off-diag function. //

<u>Proof</u> Let Skew := $\{J \in R^{n \times n}: J + J^T = 0\}$ be the set of all skew matrices and let K: $R^{n \times n} \to$ Skew be a smooth function satisfying the following condition: for all $Y \in R^{n \times n}$, Grad.off-diag.Y = -[Y,K.Y]. Let X: $L_X \subseteq R \to R^{n \times n}$ be the maximal integral curve of the gradient differential equation and let U : $L_U \subseteq R \to R^{n \times n}$ be the maximal integral

curve of the differential equation

$$U' = U(K.X), \quad U(0) = I.$$

Then $U(t)\, U(t)^\top = I$ (i.e. U maps I_U into the set $0(n)$ of orthogonal matrices) because $U(0)\, U(0)^\top = I$ and

$$(UU^\top)' = U'U^\top + U(U')^\top$$
$$= U(K.X)\, U^\top + U(U(K.X))^\top$$
$$= 0.$$

We check that the solution curve $t \to X(t)$ lies in Iso.A. In fact, we shall see that $X(t) = U(t)^\top AU(t)$. Let $Z(t) := U(t)\, X(t)\, U(t)^\top$. Then $Z(0) = A$ and

$$Z' = U'XU^\top + UX'U^\top + UX(U')^\top$$
$$= U(K.X)XU^\top + U[X,K.X]\, U^\top + UX(U(K.X))^\top$$
$$= 0.$$

Hence, for all t, $A = Z(t) = U(t)\, X(t)\, U(t)^\top$.

We now check that X and U are defined on all of R. In each case we have a differential equation of the form $Y' = f.Y$ where f is defined everywhere. Consequently, only "blow-up" can block the extension of the domain of solution. (See, for example, the book [7] by Hirsch and Smale, 1974, Chapter 8 "Fundamental Theory" Section 5 "On Extending Solutions".) We need to check that the solutions X and U remain bounded. Note that the set $0(n)$ of orthogonal matrices is a compact (i.e. closed and bounded) subset of $R^{n \times n}$. Since the continuous image of a compact set is compact, and the map $0(n) \to$ Iso.A defined by $U \to U^\top AU$ is continuous, we see that Iso.A is also compact. Thus the solutions X and U cannot blow up.

We check that the function t → off-diag.X(t) is non-increasing:

(d/dt) (t → off-diag.X(t)).t

$\qquad$ = D.off-diag.X(t).X'(t)

$\qquad$ = <∇.off-diag.X(t),

$\qquad\qquad$ -Proj.(Tan.Iso.X(t)).(∇.off-diag.X(t))>

$\qquad$ = - <Proj.(Tan.Iso.X(t)).(∇.off-diag.X(t)),

$\qquad\qquad$ Proj.(Tan.Iso.X(t)).(∇.off-diag.X(t))>

$\qquad$ ≤ 0.

Let M be an isolated relative minimum of the function off-diag. Then the function Y → off-diag.Y - off-diag.M is a strict Lapunov function for M in some neighborhood of M. (For details, see, for example the book [7] by Hirsch and Smale, 1974, Chapter 9 "On Stability of Equilibria" Section 4 "Gradient Systems".) //

The flow defined by the gradient vector field will head downhill toward equilibrium points (corresponding to t = + ∞). We turn our attention to the study of such points. We shall see that gradient lines generally head toward diagonal matrices.

<u>Proposition</u> Let X be a symmetric matrix. Then the following conditions are equivalent: Grad.off-diag.X = 0, X - Diag.X ∈ Nor.Iso.X, [X, Diag.X] = 0.

<u>Proof</u> We have the following equivalent conditions:

$$\text{Grad.off-diag.X} = 0,$$

$$\nabla.\text{off-diag.X} \in (\text{Tan.Iso.X})^{\top},$$

$$X - \text{Diag.X} \in \text{Nor.Iso.X}$$

$$[X, \text{X-Diag.X}] \text{ is symmetric.}$$

But X and Diag.X are symmetric, so [X, Diag.X] is skew. Since 0 is the only matrix that is both symmetric and skew, we have [X, Diag.X] = 0. //

<u>Corollary</u>. Let X be an n-by-n symmetric matrix. Then Grad.off-diag.X = 0 if and only if X satisfies:

$$\text{for all } i,j = 1, 2, \ldots, n, \quad x_{ij}(x_{ii} - x_{jj}) = 0.$$

<u>Proof</u> $(X(\text{Diag}.X) - (\text{Diag}.X)X)_{ij} = x_{ij}\,x_{jj} - x_{ii}\,x_{ij}.$ //

We see that if X is an equilibrium point then for all $i \neq j$, $x_{ij} = 0$ or $x_{ii} = x_{jj}$. Suppose we have for all $i \neq j$, $x_{ij} = 0$. Then X is a diagonal matrix and off-diag.X = 0. Clearly then X is a minimum point of the (always nonnegative) off-diagonal function. If the diagonal elements of X are distinct, then X is an isolated minimum.

<u>Remark</u> (Generic case) Let S be a subset of $R^{n \times n}$. We say that S is a "generic" set if S is open and dense in $R^{n \times n}$. For example, the set $\{X \in R^{n \times n}: X$ has distinct eigenvalues$\}$ is generic. While designing algorithms we are primarly interested in generic cases, because roundoff errors immediately take us out of nongeneric sets. //

The following proposition shows that equilibrium points of the gradient system that are not diagonal matrices are not stable.

<u>Proposition</u> Let A be a symmetric matrix and let X be an equilibrium point of the gradient system associated with A. Then either (i) X is a diagonal matrix and a minimum of the off-diag function on Iso.A, or (ii) X is not a diagonal matrix and not a local minimum of the off-diag function on Iso.A.

<u>Proof</u> We have already proved part (i). We now prove part (ii). Let $t \to G(t)$ be a smooth one-parameter family of orthogonal matrices satisfying $G(0) = I$. We set $Y := G(t)^{\top} XG(t)$ and $f(t) := \text{off-diag}.Y(t)$. We shall see that for an appropriate choice of G, we have $f''(0) < 0$. It follows that this choice provides a downhill direction from X. (Note $f'(0) = 0$ since X is an equilibrium point. This equation also follows from the formula for f' given below.)

Differentiating $I = G(t)\,G(t)^{\top}$, we obtain

$$0 = G'(t)\,G(t)^{\top} + G(t)\,G'(t)^{\top}\big|_{t=0}$$

$$= K + K^{\top}$$

where $K := G'(0)$, and

$$0 = G''(0)\, G(t)^T + 2G'(t)\, G'(t)^T + G(t)\, G''(t)^T |_{t=0}$$

$$= H - 2K^2 + H^T$$

where $H := G''(0)$. We shall choose $G(t)$ so that $H = K^2$. (Note that if $G(t) = \exp(Kt)$ then $G'(0) = K$ and $G''(0) = K^2$.) We also have

$$Y' = (G')^T XG + G^T XG',$$

$$Y'' = (G'')^T XG + 2(G')^T XG' + G^T XG'',$$

$$f' = \langle Y - \text{Diag}.Y,\ Y' \rangle,\ \text{and}$$

$$f'' = \langle Y' - \text{Diag}.Y',\ Y' \rangle + \langle Y - \text{Diag}.Y,\ Y'' \rangle.$$

Hence

$$Y'(0) = [X,K],\ \text{and}$$

$$Y''(0) = H^T X + 2K^T XK + XH.$$

(Also $f'(0) = \langle X - \text{Diag}.X, [X,K] \rangle = 0$.)

Using $H = K^2$, we have

$$Y''(0) = K^2 X - 2KXK + XK^2$$

$$= [[X,K],K],\ \text{and}$$

$$f''(0) = \langle [X,K] - \text{Diag}.[X,K],\ [X,K] \rangle$$

$$+ \langle X - \text{Diag}.X,\ [[X,K],K] \rangle$$

$$= \langle [X,K] - \text{Diag}.[X,K],\ [X,K] \rangle$$

$$- \langle [X - \text{Diag}.X,K],\ [X,K] \rangle$$

$$= \langle [\text{Diag}.X,K],\ [X,K] \rangle$$

$$- \langle \text{Diag}.[X,K],\ [X,K] \rangle.$$

Now assume $x_{ij} \neq 0$ where $i \neq j$. Then $x_{ii} = x_{jj}$. We take

$$K := k^{ij} \begin{bmatrix} 0 & 1 \\ -1 & 0 \end{bmatrix} \quad \text{and}\ G(t) := \exp(Kt)$$

where $K^{ij}\begin{bmatrix} a_{11} & a_{12} \\ a_{21} & a_{22} \end{bmatrix}$ is the n-by-n matrix with a_{11} in position (i,i), a_{12} in position (i,j), a_{21} in position (j,i), a_{22} in position (j,j) and 0 elsewhere. We have $G'(0) = K$ and $G''(0) = K^2$ and

$$f''(0) = <[\text{Diag}.X,K], [X,K]> - <\text{Diag}.[X,K], [X,K]>$$

$$< 0$$

because $[\text{Diag}.X,K] = 0$ and $\text{Diag}.[X,K] \neq 0$.

<u>Remark.</u> We check $[\text{Diag}.X,K] = 0$ and $\text{Diag}.[X,K] \neq 0$ when n = 3, i = 1 and j = 2:

$$[\text{Diag}.X, K] = \begin{bmatrix} x_{11} & 0 & 0 \\ 0 & x_{22} & 0 \\ 0 & 0 & x_{33} \end{bmatrix}\begin{bmatrix} 0 & 1 & 0 \\ -1 & 0 & 0 \\ 0 & 0 & 0 \end{bmatrix} - \begin{bmatrix} 0 & 1 & 0 \\ -1 & 0 & 0 \\ 0 & 0 & 0 \end{bmatrix}\begin{bmatrix} x_{11} & 0 & 0 \\ 0 & x_{22} & 0 \\ 0 & 0 & x_{33} \end{bmatrix}$$

$$= \begin{bmatrix} 0 & x_{11} & 0 \\ -x_{22} & 0 & 0 \\ 0 & 0 & 0 \end{bmatrix} - \begin{bmatrix} 0 & x_{22} & 0 \\ -x_{11} & 0 & 0 \\ 0 & 0 & 0 \end{bmatrix}$$

$$= 0$$

since $x_{11} = x_{22}$,

$$[X,K] = \begin{bmatrix} x_{11} & x_{12} & x_{13} \\ x_{12} & x_{22} & x_{23} \\ x_{13} & x_{23} & x_{33} \end{bmatrix}\begin{bmatrix} 0 & 1 & 0 \\ -1 & 0 & 0 \\ 0 & 0 & 0 \end{bmatrix} - \begin{bmatrix} 0 & 1 & 0 \\ -1 & 0 & 0 \\ 0 & 0 & 0 \end{bmatrix}\begin{bmatrix} x_{11} & x_{12} & x_{13} \\ x_{12} & x_{22} & x_{23} \\ x_{13} & x_{23} & x_{33} \end{bmatrix}$$

$$= \begin{bmatrix} -x_{12} & * & * \\ * & x_{12} & * \\ * & * & * \end{bmatrix} - \begin{bmatrix} x_{12} & * & * \\ * & -x_{12} & * \\ * & * & * \end{bmatrix}$$

$$= \begin{bmatrix} -2x_{12} & * & * \\ * & 2x_{12} & * \\ * & * & * \end{bmatrix} . \text{ //}$$

<u>Remark</u> (Alternative choice for downhill direction)

Wayne Barrett (Brigham Young University) has suggested an alternative choice for the skew-symmetric matrix K which appears in the last part of the preceeding proof; it is based on the following fact.

<u>Proposition</u> Let X be a symmetric matrix and let

$$X = \text{Lower.X} + \text{Diag.X} + (\text{Lower.X})^T$$

be the decomposition of X into its strictly lower triangular part, its diagonal part and its strictly upper triangular part. If $[\text{Diag.X},X] = 0$ and $X \neq \text{Diag.X}$ then the skew-symmetric matrix $K:= \text{Lower.X} -(\text{Lower.X})^T$ satisfies

$$[\text{Diag.X},K] = 0 \text{ and } \text{Diag.}[X,K] \neq 0.$$

<u>Proof</u> We have

$$0 = [\text{Diag.X},X]$$
$$= [\text{Diag.X}, \text{Lower.X}] + [\text{Diag.X}, \text{Diag.X}] + [\text{Diag.X}, (\text{Lower.X})^T]$$
$$= [\text{Diag.X}, \text{Lower.X}] + [\text{Diag.X}, (\text{Lower.X})^T].$$

Since $[\text{Diag.X}, \text{Lower.X}]$ is strictly lower triangular and $[\text{Diag.X}, (\text{Lower.X})^T]$ is strictly upper triangular, we have

$$0 = [\text{Diag.X}, \text{Lower.X}] = [\text{Diag.X}, (\text{Lower.X})^T]$$

and hence

$$0 = [\text{Diag.X},K].$$

We also have

$$[X,K] = [\text{Lower.X} + \text{Diag.X} + (\text{Lower.X})^T, K]$$
$$= [\text{Lower.X} + (\text{Lower.X})^T, \text{Lower.X} - (\text{Lower.X})^T] + [\text{Diag.X},K]$$
$$= [(\text{Lower.X})^T, \text{Lower.X}] - [\text{Lower.X}, (\text{Lower.X})^T]$$
$$= 2 [(\text{Lower.X})^T, \text{Lower.X}].$$

We complete the proof by applying the following lemma.

<u>Lemma</u> Let L be a strictly lower triangular matrix. If $[L, L^T] = 0$ then $L = 0$.

<u>Proof</u> We proceed by induction on the size of L. For n = 2, we have

$$[L, L^T] = \begin{bmatrix} 0 & 0 \\ a & 0 \end{bmatrix} \begin{bmatrix} 0 & a \\ 0 & 0 \end{bmatrix} - \begin{bmatrix} 0 & a \\ 0 & 0 \end{bmatrix} \begin{bmatrix} 0 & 0 \\ a & 0 \end{bmatrix}$$

$$= \begin{bmatrix} -a^2 & 0 \\ 0 & a^2 \end{bmatrix} .$$

For the induction step, we partition L as follows:

$$L =: \begin{bmatrix} 0 & 0 \\ m & M \end{bmatrix}$$

where $m \in R^{n-1}$ and $M \in R^{n-1 \times n-1}$. From

$$0 = [L, L^T]$$

$$= \begin{bmatrix} 0 & 0 \\ m & M \end{bmatrix} \begin{bmatrix} 0 & m^T \\ 0 & M \end{bmatrix} - \begin{bmatrix} 0 & m^T \\ 0 & M \end{bmatrix} \begin{bmatrix} 0 & 0 \\ m & M \end{bmatrix}$$

$$= \begin{bmatrix} -m^T m & * \\ * & * \end{bmatrix}$$

We obtain m = 0. Hence

$$0 = [L, L^T] = \begin{bmatrix} 0 & 0 \\ 0 & MM^T - M^T M \end{bmatrix}$$

and we apply the induction hypothesis to conclude that M = 0. //

<u>An Isospectral Gradient Flow for Complex Hermitian Matrices</u>

In this section, I shall consider the following ("spectral") problem:
Given a complex Hermitian matrix, find its eigenvalues. I shall use the
theory of ordinary differential equations to solve this problem. In
particular, I shall describe an isospectral dynamical system on Hermitian

matrices that flows "downhill" toward diagonal matrices. The choice of the system is based on the following idea: Look for a minimum of a real-valued function that measures the size of the off-diagonal part of a matrix. The results in this section are similar to those which appear in the section on real symmetric matrices. Consequently, I omit (or only sketch) many of the proofs.

Let $\mathbb{C}$ denote the set of complex numbers and let $\mathbb{C}^{n \times n}$ denote the set of all n-by-n matrices over $\mathbb{C}$. Recall that the ("Frobenius") inner product $< , >: \mathbb{C}^{n \times n} \times \mathbb{C}^{n \times n} \to R$ is defined by $<X,Y>: = \text{trace } (XY^*)$. (I use a superscript $*$ to denote the conjugate-transpose operation. Note $<X,Y> = \sum x_{ij} \bar{y}_{ij}$.) Define the function

$$\text{off-diag}: \mathbb{C}^{n \times n} \to R \quad \text{by} \quad X \to <X - \text{Diag}.X, X - \text{Diag}.X>,$$

where $\text{Diag}.X := \text{Diag } (x_{11}, x_{22}, \ldots, x_{nn})$.

<u>Proposition</u> (Derivative) The derivative of the off-diag function is given by

$$\text{D.off-diag}.X.H = <\nabla.\text{off-diag}.X, H>$$

where $\nabla.\text{off-diag}.X := X - \text{Diag}.X$.

We shall call $\nabla.\text{off-diag}.X$ the "spatial gradient" of the off-diag function at X. We define the isospectral surface determined by a matrix X as follows:

$$\text{Iso}.X := \{U^*XU : U \in \mathbb{C}^{n \times n} \text{ and } U^*U = I\}.$$

<u>Remark</u> Let $U(n) := \{U \subset \mathbb{C}^{n \times n}: U^*U = I\}$ be the unitary group and let $\text{Herm} := \{X \in \mathbb{C}^{n \times n}: X^* = X\}$ be the set of Hermitian matrices. Define the group action $*: U(n) \times \text{Herm} \to \text{Herm}$ by $U^*I := U^*XU$. Then Iso.X is the orbit of X. //

<u>Proposition</u> Let X be an n-by-n complex matrix. Then the tangent space at X to the surface Iso. X is

$$\text{Tan.Iso}.X := \{[X,K]: K \in \mathbb{C}^{n \times n} \text{ and } K + K^* = 0\}$$

where $[X,K] := XK - KX$.

Remark If K + K* = 0 we say that K is "skew" (or "skew-Hermitian"). Note that if X is Hermitian then the matrix

$$[X,K] = XK - KX = XK + (XK)*$$

is also Hermitian.

Let K = Re.K + i Im.K be the decomposition of $K \in \mathbb{C}^{n \times n}$ into its real and imaginary parts. Note that K is skew-Hermitian if and only if Re.K is skew-symmetric and Im.K is symmetric. Consequently, the skew Hermitian matrices form a vector space over R with dimension equal to n^2. //

For $X \in \mathbb{C}^{n \times n}$ we set

$$\text{Nor.Iso.X} := \{Y \in \mathbb{C}^{n \times n}: X*Y = YX*\}.$$

We call this set the "normal space" at X to the surface Iso.X.

Proposition Let X be an n-by-n matrix. Then

$$\text{Nor.Iso.X} = (\text{Tan.Iso.X})^{\perp}.$$

Lemma (Adjoint of X → [A,X]) Let A,X and Y be n-by-n matrices. Then

$$\langle [A,X], Y \rangle = \langle X, [A*,Y] \rangle.$$

Lemma Let Y be an n-by-n matrix. If Y satisfies the following condition

$$\text{for all skew K, } \langle K, Y \rangle = 0$$

then Y = 0.

Proof For $p \neq q$ let K^{pq} be the skew matrix with 1 in position (p,q), -1 in position (q,p) and 0 elsewhere. Then $\langle Y, K^{pq} \rangle = Y_{pq} - Y_{qp}$. For any pair (p,q) ($p = q$ allowed) let S^{pq} be the symmetric matrix with 1 in position (p,q), 1 in position (q,p) and 0 elsewhere. Then iS^{pq} is skew and $\langle Y, iS^{pq} \rangle = -i(Y_{pq} + Y_{qp})$. If $p \neq q$ then, from $0 = Y_{pq} - Y_{qp}$ and $0 = Y_{pq} + Y_{qp}$ we get $0 = Y_{pq} = Y_{qp}$. If $p = q$ then, from $0 = 2 Y_{pp}$, we get $Y_{pp} = 0$. //

<u>Proof</u> (of proposition) We have the following equivalent conditions:

$$Y \in (\text{Tan. Iso. } X)^{\perp},$$

$$\text{for all skew } K, \quad 0 = <[X,K],Y>,$$

$$\text{for all skew } K, \quad 0 = <K,[X^*,Y]>,$$

$$0 = [X^*,Y]. \quad //$$

We define the "(isospectral) surface gradient" of the function off-diag at X as the projection onto the tangent space of the spatial gradient:

$$\text{Grad.off-diag.} X := \text{Projection.}(\text{Tan.Iso.}X).(\nabla.\text{off-diag.}X).$$

<u>Proposition</u> (Isospectral Gradient Flow) Let A be a Hermitian matrix. Then there exists a unique smooth function $X : R \to \text{Iso. } A$ satisfying

$$X' = -\text{Grad. off-diag. } X$$

$$X(0) = A.$$

The function off-diag is nonincreasing along X. If M is an isolated minimum on Iso.A of the function off-diag then M is an asymptotically stable equilibrium point of this gradient system.

We now consider the equilibrium points of the given gradient system.

<u>Proposition</u> Let X be a Hermitian matrix. Then

$$\text{Grad.off-diag.} X = 0 \text{ if and only if } [X, \text{Diag.}X] = 0.$$

<u>Proof</u> We have the following equivalent conditions:

$$\text{Grad.off-diag.} X = 0,$$

$$\nabla.\text{off-diag.}X \in (\text{Tan.Iso.}X)^{\perp}$$

$$X - \text{Diag.}X \in \text{Nor.Iso.}X$$

$$[X, X - \text{Diag.}X] = 0. \quad //$$

<u>Corollary</u> Let X be an n-by-n Hermitian matrix. Then $\text{Grad.off-diag.}X = 0$ if and only if X satisfies:

$$\text{for all } p,q = 1,2, \ldots, n, \quad x_{pq}(x_{pp} - x_{qq}) = 0.$$

We see that if Hermitian matrix X is an equilibrium point then for all $p \neq q$,

$$x_{pq} = 0 \text{ or } x_{pp} = x_{qq}.$$

Suppose we have, for all $p \neq q$, $x_{pq} = 0$. Then X is a diagonal matrix and off-diag.X = 0. Clearly then X is a minimum point of the off-diag function. If the diagonal elements of X are distinct then X is an isolated minimum. The following proposition shows that the equilibrium points that are not diagonal matrices are not stable.

<u>Proposition</u> Let X be an n-by-n Hermitian matrix satisfying Grad. off-diag. X = 0. If X is not a diagonal matrix, then X is not a local minimum on Iso.X of the off-diag function.

<u>Proof</u> Let $t \to G(t)$ be a smooth one parameter family of unitary matrices satisfying G(0) = I. We set Y(t) := G(t)*XG(t) and f(t) := off-diag.Y(t). We shall see that for an appropriate choice of G(t) we have f''(0) < 0. Differentiating I = G(t)G(t)* we obtain

$$0 = K + K*$$

where K := G'(0), and

$$0 = H + H* - 2K^2$$

where H := G''(0). We shall choose G(t) so that $H = K^2$. We also have

$$Y'(0) = [X,K],$$

$$Y''(0) = H*X + 2K^T XK + XH$$

$$= [[X,K],K], \text{ and}$$

$$f''(0) = <[Diag.X,K], [X,K]> - <Diag.[X,K], [X,K]>.$$

The proof now splits into two cases.

<u>Case</u> Re.$x_{pq} \neq 0$. We take

$$G(t) := \exp(Kt),$$

where

$$K := K^{pq} \begin{bmatrix} 0 & 1 \\ -1 & 0 \end{bmatrix}.$$

We have

$$f''(0) < 0$$

because $[\mathrm{Diag}.X,K] = 0$ and $\mathrm{Diag}.[X,K]_{pp} = -2\,\mathrm{Re}.x_{pq}$.

<u>Remark</u> We check this last assertion when $n = 2$, $p = 1$ and $q = 2$:

$$[X,\ K] = \begin{bmatrix} x_{11} & x_{12} \\ x_{21} & x_{22} \end{bmatrix} \begin{bmatrix} 0 & 1 \\ -1 & 0 \end{bmatrix} - \begin{bmatrix} 0 & 1 \\ -1 & 0 \end{bmatrix} \begin{bmatrix} x_{11} & x_{12} \\ x_{21} & x_{22} \end{bmatrix}$$

$$= \begin{bmatrix} -x_{12} & x_{11} \\ -x_{22} & x_{21} \end{bmatrix} - \begin{bmatrix} x_{21} & x_{22} \\ -x_{11} & -x_{12} \end{bmatrix}$$

$$= \begin{bmatrix} -2\,\mathrm{Re}.x_{12} & 0 \\ 0 & 2\,\mathrm{Re}.x_{12} \end{bmatrix}. \ //$$

<u>Case</u> $\mathrm{Im}.x_{pq} \neq 0$. We take

$$G(t) := \exp(Kt),$$

where

$$K := K^{pq} \begin{bmatrix} 0 & i \\ i & 0 \end{bmatrix}.$$

We have $f''(0) < 0$ because $[\mathrm{Diag}.X,K] = 0$ and $\mathrm{Diag}.\ [X,K]_{pp} = 2\,\mathrm{Im}.x_{pq}$.

<u>Remark</u> We check this last assertion when $n = 2$, $p = 1$ and $q = 2$:

$$[X,\ K] = \begin{bmatrix} x_{11} & x_{12} \\ x_{21} & x_{22} \end{bmatrix} \begin{bmatrix} 0 & i \\ i & 0 \end{bmatrix} - \begin{bmatrix} 0 & i \\ i & 0 \end{bmatrix} \begin{bmatrix} x_{11} & x_{12} \\ x_{21} & x_{22} \end{bmatrix}$$

$$= i \begin{bmatrix} x_{12} & x_{11} \\ x_{22} & x_{21} \end{bmatrix} - i \begin{bmatrix} x_{21} & x_{22} \\ x_{11} & x_{12} \end{bmatrix}$$

$$= \begin{bmatrix} 2 \, \text{Im}.x_{12} & 0 \\ 0 & -2 \, \text{Im}.x_{12} \end{bmatrix} . \; //$$

Concluding Remarks

In this paper I described an isospectral gradient differential equation which may be used to determine the eigenvalues of a symmetric (or Hermitian) matrix. This gradient differential equation can be used as the basis for a "gradient" numerical algorithm. Questions concerning efficiency of this algorithm then arise: What is the rate of convergence of this gradient algorithm? (Compare Dieft, Nanda and Tomei [4].) What is the "total" cost for this gradient algorithm? (See Smale [12].) Is this gradient algorithm competitive with other algorithms? I have not tried to answer these questions. I am more interested in trying to adapt these gradient algorithms to problems in which the matrix has special structure (e.g. when the matrix is Toeplitz).

In the references, I have listed several books that I have not yet cited. They are included to give the reader a place to start in the library. Most of the little that I know about differential geometry and Lie groups I learned from the 1984 book [15] by Thorpe and the 1979 book [2] by Curtis. The eigenproblem is understood best (and calculated fastest) for circulant matrices; the delightful book [3] by P. J. Davis is a good introduction to them. The papers [4, 10, 16] by Dieft, Nanda, Tomei and Watkins strongly influenced the content of this paper; in particular, I adapted some of their proofs.

<u>References</u>

1. M. T. Chu, The generalized Toda flow, the QR algorithm and the center manifold theory, SIAM J. Alg. Disc. Math. 5 (1984) pp. 187-201.

2. Curtis, M. L., Matrix Groups, Springer, 1979.

3. Davis, P. J., Circulant Matrices, Wiley, 1979.

4. P. Dieft, T. Nanda and C. Tomei, Ordinary differential equations and the symmetric eigenvalue problem, SIAM J. Numerical Analysis, 20, (1983) pp. 1-22.

5. H. Flascka, The Toda lattice I, Phys. Rev. B 9 (1974) pp. 1924-1925.

6. Golub, G. and Van Loan, C. F., Matrix Computations, Johns Hopkins University Press, 1983.

7. Hirsch, M. W. and Smale, S., Differential Equations, Dynamical Systems, and Linear Algebra, Academic, 1974.

8. S. Lang, Real Analysis, Addison-Wesley, 1969.

9. J. Moser, Finitely many mass points on the line under the influence of an exponential potential - an integrable system, in the book: Dynamical Systems Theory and Applications, edited by J. Moser, Lecture Notes in Physics, 38, Springer, 1975, pp. 467-497.

10. T. Nanda, Differential equations and the QR algorithm, SIAM J. Numerical Analysis 22 (1983) pp. 310-321.

11. Parlett, B. N., The Symmetric Eigenvalue Problem, Prentice-Hall, 1980.

12. S. Smale, On the efficiency of algorithms, AMS Bull. 13 (1985) pp. 87-121.

13. W. W. Symes, Hamiltonian group actions and integrable systems, Physica 1 D (1980), pp. 339-374.

14. W. W. Symes, The QR algorithm and scattering for the finite nonperiodic Toda lattice, Physica 4D (1982), pp. 272-280.

15. Thorpe, J. A., Elementary Topics in Differential Geometry, Springer, 1979.

16. D. S. Watkins, Isospectral flows, SIAM Review 26 (1984) pp. 379-391.

Acknowledgements

I wrote this paper during 1986 while I visited the following institutions: Rice University, the University of Wyoming, the Mathematisches Forschungsinstitute Oberwolfach and Brigham Young University. I wish to thank the people who made these pleasant visits possible. While doing this research I received some financial support from the Mathematisches Forschungsinstitute (arranged by John C. Cannon of Washington State University and Ulrich Hornung of University Bundeswehr Munchen), from the J. E. Warren Distinguished Lectureship Fund (arranged by Eli Isaacson and Richard Ewing of the University of Wyoming) and from the Brigham Young Visiting Lectureship Fund (arranged by Wayne Barrett and Peter Crawley). I also wish to thank Richard McCullogh (Amoco Production Company, Houston) and William W. Symes who made possible my visit to Rice University. This paper is an expanded version of talks I gave in 1986: in February, at the meeting of the Texas-Oklahoma section of SIAM at the University of Tulsa; in April, at the University of Wyoming; in May, at the Mathematisches Forschungsinstitute; in June, at Brigham Young University; and, in August, at the International Congress of Mathematicians at the University of California in Berkeley.

In addition to the people mentioned above, I wish to thank colleagues who listened to me talk about isospectral flows and offered encouragement (and I apologize to those I have forgotten to list): Ray Head, John Scales, Sam Gray, Dan Vasicek (all from Amoco Production Company, Research Center, Tulsa), Morton Curtis (Rice University), Phillip E. Parker (Wichita State University), Margaret Cheney (Duke University), and David Sattinger (University of Minnesota). Wayne Barrett (Brigham Young University) critically read a draft of this paper. Eli Isaacson (University of Wisconsin) made suggestions which led to substantial simplifications of some of the proofs. I also wish to thank Jill Fielding (Brigham Young University) for her careful typing of most of this paper and Lonette Stoddard (Brigham Young University) for her typing of the introduction and bibliography.

Kenneth R. Driessel
c/o G. Landman, Attorney
1921 South Boston Avenue
Tulsa, OK 74119

International Series of
Numerical Mathematics, Vol. 77

ON AN INTEGRAL EQUATION OF THE FIRST KIND IN INVERSE ACOUSTIC SCATTERING

Andreas Kirsch and Rainer Kress

Abstract: The inverse problem under consideration is to determine the shape of a scatterer from the far-field pattern of the scattered wave for an incident time-harmonic plane wave. A new method for approximately solving this problem is presented.

1   Introduction

The inverse scattering problem for acoustic and electromagnetic waves, which consists in recovering the shape of a scatterer from the scattered field, has received much attention in various areas such as remote sensing, nondestructive testing, ultra sound medicine, seismic imaging etc. For a survey on recent progress in inverse time-harmonic acoustic scattering for low and intermediate values of the wave number we refer to the expository papers by Colton [2] and Sleeman [13] and the monograph by Colton and Kress [3].

The inverse acoustic scattering problem is difficult to solve due to two main reasons: it is nonlinear and improperly posed. Of these two, it is the latter that presents the more basic difficulty. Roughly speaking inverse scattering is improperly posed since it involves some kind of analytic continuation of the scattered field and it is nonlinear since finding the shape of the obstacle means locating level surfaces for the scattered field. For the analytic continuation we propose seeking the scattered field in the form of an acoustic single-layer potential on some surface within the unknown scatterer. This leads to an ill-posed linear integral equation of the first kind which is stabilized through Tikhonov regularization. Then for determining the shape of the scatterer which we assume to be sound soft we minimize the L^2-defect of the total field over a suitably chosen family of admissible surfaces.

To motivate our approach in Section 2 we first describe the approximate solution of the corresponding direct scattering problem with the use of internal acoustic single-layer potentials via an ill-posed integral equation of the first kind. Then in Section 3 we give a detailed discussion of our method for the inverse problem.

2 The direct scattering problem

In order to describe our approach we first need to give the formulation of the exterior boundary–value problem of time–harmonic acoustic scattering by a sound soft obstacle. Let $D \subset \mathbb{R}^m$, $m = 2, 3$, be a bounded domain with a sufficiently smooth connected boundary surface ∂D and let an incident wave u^i be given by the plane wave $u^i(x) = exp[ik(x, \alpha)]$, $x \in \mathbb{R}^m$, where $k > 0$ is the wave number and where the unit vector α represents the direction of the incident wave. If we denote the scattered wave by u^s and define the total wave by $u = u^i + u^s$, then the direct scattering problem for the sound soft obstacle D consists in finding a solution $u \in C^2(\mathbb{R}^m \setminus \overline{D}) \cap C(\mathbb{R}^m \setminus D)$ to the Helmholtz equation

$$(2.1) \qquad \Delta u + k^2 u = 0 \text{ in } \mathbb{R}^m \setminus \overline{D}$$

with homogeneous Dirichlet boundary condition

$$(2.2) \qquad u = 0 \text{ on } \partial D$$

such that the scattered wave u^s satisfies the Sommerfeld radiation condition

$$(2.3) \qquad \frac{\partial u^s}{\partial r} - iku^s = o\left(r^{\frac{1-m}{2}}\right), \ r = |x| \to \infty,$$

uniformly for all directions.

The radiation condition ensures the uniqueness for this exterior Dirichlet problem by Rellich's lemma. The classical existence proofs due to Vekua, Weyl and Müller reduce the boundary–value problem to a boundary integral equation of the second kind by seeking the solution in the form of an acoustic double–layer potential with a density distributed over the boundary surface ∂D. For a detailed study of this boundary integral equation method we refer to Colton and Kress [3] for classical solutions and Kirsch [6] for weak solutions.

Here we describe a different method. We choose an auxiliary smooth surface Γ contained in D and try to find the solution to the exterior Dirichlet problem in the form of an acoustic single–layer potential

$$(2.4) \qquad u^s(x) = \int_\Gamma \Phi(x, y)\varphi(y)ds(y), \ x \in \mathbb{R}^3 \setminus D,$$

with a source density φ distributed over Γ. Here Φ denotes the free–space fundamental solution to the Helmholtz equation, this means

$$\Phi(x, y) = \frac{1}{4\pi} \frac{e^{ik|x-y|}}{|x-y|}, \ x \neq y, \text{ in } \mathbb{R}^3$$

and

$$\Phi(x, y) = \frac{i}{4} H_0^{(1)}(k|x-y|), \ x \neq y, \text{ in } \mathbb{R}^2.$$

The potential (2.4) solves the exterior Dirichlet problem (2.1) to (2.3) provided the density φ is a solution of the integral equation of the first kind

$$(2.5) \qquad \int_\Gamma \Phi(x, y)\varphi(y)ds(y) = -u^i(x), \ x \in \partial D.$$

We introduce an integral operator $S : L^2(\Gamma) \to L^2(\partial D)$ by

$$(2.6) \qquad (S\varphi)(x) := \int_\Gamma \Phi(x,y)\varphi(y)ds(y), \quad x \in \partial D,$$

and rewrite the integral equation (2.5) in the form

$$(2.5') \qquad S\varphi = -u^i.$$

Since the boundary ∂D and the auxiliary surface Γ are disjoint the integral operator S has a smooth kernel and therefore it is compact and cannot have a bounded inverse. Hence, the integral equation (2.5) is ill–posed.

We can assure uniqueness by choosing Γ such that the wave number k is not an interior Dirichlet eigenvalue for the interior of Γ, this means the homogeneous Dirichlet problem in the interior of Γ admits only the trivial solution. Then, by standard arguments, it can be seen that S is injective.

Unfortunately, in general, we cannot expect that a solution to equation (2.5) exists. Since the single–layer potential (2.4) is an analytic function in the exterior of Γ, equation (2.5) can have a solution only for those incoming waves u^i for which the scattered wave u^s can be analytically extended into the exterior of Γ. The answer to this question is related to the famous Rayleigh hypothesis (c.f. Millar [10] or van den Berg and Fokkema [1]). For example, for an ellipse the scattered wave u^s for plane incident waves can be analytically extended into any domain not containing the straight line connecting the two foci of the ellipse. However, for an arbitrary domain there is no criterion available to decide whether and how far the scattered wave can be extended analytically across the boundary surface ∂D.

One of the basic techniques to treat ill–posed integral equations of the first kind is the classical Tikhonov regularization which seeks approximate solutions φ_α to (2.5) by solving

$$(2.7) \qquad \alpha\varphi_\alpha + S^*S\varphi_\alpha + S^*u^i = 0$$

where $\alpha > 0$ denotes a suitably chosen regularization paramater and where $S^* : L^2(\partial D) \to L^2(\Gamma)$ denotes the adjoint of S, given by

$$(2.8) \qquad (S^*g)(x) := \int_{\partial D} \overline{\Phi(x,y)}g(y)ds(y).$$

In general α will be chosen close to zero so that the uniquely solvable equation (2.7) of the second kind and its discretized version is still reasonably well conditioned while the error due to the introduction of α is kept small.

Note, that solving the regularized equation (2.7) is equivalent to minimizing the Tikhonov functional

$$(2.9) \qquad \|S\varphi_\alpha + u^i\|^2_{L^2(\partial D)} + \alpha\|\varphi_\alpha\|^2_{L^2(\Gamma)},$$

representing the L^2–defect of $S\varphi_\alpha + u^i$ on the boundary penalized by the L^2–norm of φ_α. This observation shows that the single–layer potential on the internal surface Γ may be considered as limiting case of the boundary defect minimizing approach using internal sources as described among others by Kirsch [6], Limic [9] and Müller and Kersten [12]. Here one chooses sources $\varphi_1, ..., \varphi_n$ at a finite number of internal points $y_1, ..., y_n \in D$ such that the defect on the boundary

$$\left\| \sum_{j=1}^{n} \Phi(.,y_j)\varphi_j + u^i \right\|_{L^2(\partial D)}$$

is minimized. For similar reasons as described above this approach becomes ill–posed as the number n of internal source points increases, and indeed, corresponding numerical instabilities have been reported by Limic [9].

As is always the case with Tikhonov regularization, we can expect convergence of the solution φ_α to (2.7) as $\alpha \to 0$ only in the case where the original equation (2.5) has a solution. But fortunately, in order to obtain approximate solutions to the exterior scattering problem by

$$(2.10) \qquad u_\alpha^s(x) = \int_\Gamma \Phi(x,y)\varphi_\alpha(y)ds(y), \ x \in \mathbb{R}^m \setminus D$$

we do not need convergence of the source densities φ_α. Due to the well–posedness of the exterior Dirichlet problem small deviations $\|u_\alpha^s - u_s\|_{L^2(\partial D)}$ on the boundary ensure small deviations $\|u_\alpha^s - u_s\|_{H^{1/2}_{loc}(\mathbb{R}^m \setminus \overline{D})}$ in the solution all over the domain $\mathbb{R}^m \setminus \overline{D}$. Therefore, what we actually want is convergence $\|S\varphi_\alpha + u^i\|_{L^2(\partial D)} \to 0$, $\alpha \to 0$, in order to satisfy the boundary condition (2.2) approximately.

Using familiar techniques from the theory of Tikhonov regularization the following convergence can be established (see Kress and Mohsen [7]).

Theorem 2.1 Assume that k is not a Dirichlet eigenvalue for the interior of Γ. Then

$$\|S\varphi_\alpha + u^i\|_{L^2(\partial D)} \to 0, \ \alpha \to 0.$$

Provided $S\varphi_\alpha + u^i = 0$ is solvable, then

$$\|S\varphi_\alpha + u^i\|_{L^2(\partial D)} = 0(\sqrt{\alpha}), \ \alpha \to 0.$$

The condition $u^i \in SS^*(L^2(\partial D))$ is necessary and sufficient for the optimal convergence rate

$$\|S\varphi_\alpha + u^i\|_{L^2(\partial D)} = 0(\alpha), \ \alpha \to 0.$$

This method using an internal single–layer potential is simple since as apposed to the boundary integral equations it leads to smooth kernels which allow the application of basic quadrature rules without any need to deal with singular integrals. In addition, the nonuniqueness problems which cause difficulties in the boundary integral equations, can be easily avoided. For a numerical example see Kress and Mohsen [7]. Despite this advantage, since in general we may have only relatively poor convergence for the regularization parameter α tending to zero, this procedure certainly may only serve as an illustrative example for the Tikhonov regularization technique but will not become a serious competitor for the well established boundary integral equation techniques. However, it may turn out very useful for the treatment of inverse problems as described in the next section.

3   The inverse problem

From the radiation condition (2.3) it can be deduced (see Colton and Kress [3]) that the scattered wave u^s has the asymptotic expansion

$$(3.1) \qquad u^s(x) = \frac{e^{ikr}}{r^{\frac{m-1}{2}}} \left[f(\hat{x}) + 0\left(\frac{1}{r}\right) \right]$$

where $\hat{x} := x/|x|$ and where f is known as the far–field pattern of the scattered wave. The inverse scattering problem we want to consider is, given the far–field pattern f for one or more incident plane waves, to find the shape of the scatterer D.

Since solutions to the Helmholtz equation are analytic the far–field pattern is an analytic function on the unit sphere Ω in $\mathbb{R}^m$. Furthermore, due to a result by Müller [11], in the case $m = 3$, the coefficients in the expansion with respect to orthonormalized spherical harmonics

$$f(\hat{x}) = \sum_{n=0}^{\infty} \sum_{l=-n}^{n} a_{nl} Y_n^l(\hat{x}), \hat{x} \in \Omega,$$

satisfy a growth condition of the form

$$(3.2) \qquad \sum_{n=0}^{\infty} \left(\frac{2n}{keR} \right)^{2n} \sum_{l=-n}^{n} |a_{nl}|^2 < \infty$$

for all R such that $\overline{D}$ is contained in the open ball of radius R and center at the origin. A similar result holds in the case $m = 2$. From this regularity of the far–field pattern it is obvious that the inverse scattering problem is severely ill–posed. In addition, it is nonlinear since the far–field depends nonlinearly on the boundary surface ∂D.

For the question of uniqueness to the inverse scattering problem we refer to Colton and Sleeman [4] and Jones [5]. In particular, due to Schiffer, D is uniquely determined by a knowledge of the far–field pattern f on some surface patch of the unit sphere for one incoming wave and for an interval of wave numbers k. In an alternative version D is uniquely determined by the far–field patterns for some finite number of incident waves of the same wavenumber k with different directions α.

We propose the following method to approximately solve the inverse scattering problem. In a first step, motivated by the approach for the direct problem described in the previous section, we deal with the ill–posed linear part of the inverse problem by trying to analytically extend the far–field pattern into the exterior of the unknown scatterer in the form of a single–layer potential on some internal surface Γ. For this we need to have some a–priori knowledge let say on the size of a ball contained in D. The single–layer potential

$$(3.3) \qquad u_s(x) = \int_{\Gamma} \Phi(x,y)\varphi(y)ds(y), \quad x \in \mathbb{R}^m \setminus \Gamma$$

has the asymptotic expansion

$$u_s(x) = c_m \frac{e^{ikr}}{r^{\frac{m-1}{2}}} \left[\int_{\Gamma} e^{-ik(\hat{x},y)}\varphi(y)ds(y) + 0\left(\frac{1}{r}\right) \right], \quad |x| \to \infty,$$

where

$$c_m := \begin{cases} \frac{1}{4\pi} & , m = 3, \\[2mm] \frac{i}{4}\, e^{-\frac{i\pi}{4}}\, \sqrt{\frac{2}{\pi k}} & , m = 2. \end{cases}$$

We introduce the far-field operator $F : L^2(\Gamma) \to L^2(\Omega)$ by

$$(3.4) \qquad (F\varphi)(\hat{x}) := \int_\Gamma e^{-ik(\hat{x},y)} \varphi(y) ds(y), \ \hat{x} \in \Omega.$$

Then, given the far-field pattern f, we have to solve the integral equation of the first kind

$$(3.5) \qquad F\varphi = \tilde{f}$$

where $\tilde{f} := f/c_m$. Again, we treat this ill-posed equation by Tikhonov regularization

$$(3.6) \qquad \alpha\varphi_\alpha + F^* F\varphi_\alpha = F^* f$$

where $F^* : L^2(\Omega) \to L^2(\Gamma)$ denotes the adjoint of F, given by

$$(3.7) \qquad (F^*\psi)(y) := \int_\Omega e^{ik(\hat{x},y)} \psi(\hat{x}) ds(\hat{x}), \ y \in \Gamma.$$

We may consider this procedure as limiting case of the direct problem with boundary data given on some large sphere. Here, of course, we want to extend the solution into the interior of this sphere and therefore we need convergence of the densities φ_α as $\alpha \to 0$ rather then only convergence of $F\varphi_\alpha \to \tilde{f}$ as $\alpha \to 0$. This convergence is true only if the equation (3.5) is solvable, that means, if the scattered wave belonging to f can be continued analytically into the exterior of Γ. Therefore the success of our approach hinges also on the Rayleigh hypothesis.

After we have determined φ_α and the corresponding approximation u_α^s for the scattered wave u^s, in the second step, we deal with the nonlinear part of the inverse scattering problem and seek the unknown surface as the location of the zeros of $u^s + u^i$. For this, we suggest to select some compact family M of surfaces described through a convenient parameterization and possibly include some constraints corresponding to a-priori informations on the unknown scatterer ∂D. Then we seek an approximation to ∂D by minimizing

$$(3.8) \qquad \int_A |u_\alpha^s + u^i|^2 ds$$

over all surfaces A in M. If we use the far-field for more than one incident wave we have to minimize the corresponding sum over all waves.

In a first example we try to recover an ellipse from its far-field pattern. We have used as data the approximate far-field pattern produced by the classical boundary integral equation method for the direct problem. For the numerical solution of the integral equation we applied Nyström's quadrature method with 64 knots using a numerical integration which takes proper care of the logarithmic singularities of the kernels (see Kussmaul [8]).

The integral equation (3.6) again is solved through Nyström's method using the trapezoidal rule with 32 knots. In our numerical experiment we chose the regularization parameter such that the L^2-defect of $u^s_\alpha + u^i$ becomes minimal on the correct boundary. But we found the reconstruction of the same accuracy for a large range of α varying from 10^{-6} to 10^{-12}.

In a preliminary version, for the second step, instead of using the full optimization problem (3.8) we minimized $|u^s_\alpha + u^i|$ along a finite number – actually 16 – of radial rays – equidistantly spaced – going out from the internal curve Γ.

For the internal curve Γ we again choose ellipses: one reasonably close and the second further away from the boundary to recover. From the figures it is obvious that the choice of the internal curve Γ is crucial for the quality of the reconstruction. They also make clear that our approach needs to be implemented interactively: firstly, make some initial guess for an appropriate internal curve Γ, based on some a-priori knowledge on the scatterer D; secondly, improve the internal curve Γ by inspection of the reconstruction obtained through this choice of Γ and then iterate this procedure.

Our results also indicate that increasing the number of incident waves increases the accuracy of the reconstruction. In our examples we have used either one or two incoming plane waves with directions indicated through the arrows in the figures.

In all cases the ellipse to reconstruct has semiaxis $a = 1$ and $b = 1.5$ and the wavenumber used is $k = 1$. For Fig.1a and 1b the internal ellipse has semiaxis $a' = 0.6$ and $b' = 1.2$, for Fig.1c and 1d we have $a' = 0.6$ and $b' = 0.6$. In Fig.1a and 1c one incident field is used with directions $\alpha = 0$, and in Fig.1b and 1d two incident fields are used with direction $\alpha = 0$ and $\alpha = \pi/2$. Fig.1a to 1d give the reconstruction from data without error whereas Fig.2a to 2d give the corresponding results for inexact data with a noise level of 5 %.

Further research will have to incorporate the strong regularity of the far-field pattern into the Tikhonov regularization by choosing some norm on Γ reflecting the property (3.2), in particular in the case of noisy far-field data. Finally, we also will have to combine the two steps (3.6) and (3.8) by minimizing simultaneously

$$(3.9) \qquad \|F\varphi_\alpha + u^i\|^2_{L^2(\Omega)} + \alpha\|\varphi_\alpha\|^2_{L^2(\Gamma)} + \|u^s_\alpha + u^i\|^2_{L^2(A)}$$

over all $\varphi_\alpha \in L^2(\Gamma)$ and $A \in \mathcal{M}$ in order to guarantee some convergence as $\alpha \to 0$ in the case where $F\varphi = f$ is not solvable.

4 References

1. van den Berg, P.M and Fokkema, J.T.: The Rayleigh hypothesis in the theory of diffraction by a cylindrical obstacle. IEEE Trans. Antennas and Prop. AP-27 (1979), 577–583.

2. Colton, D.: The inverse scattering problem for time-harmonic acoustic waves. SIAM Review 26 (1984), 323–350.

3. Colton, D. and Kress, R.: Integral Equation Methods in Scattering Theory, John Wiley, New York, 1983.

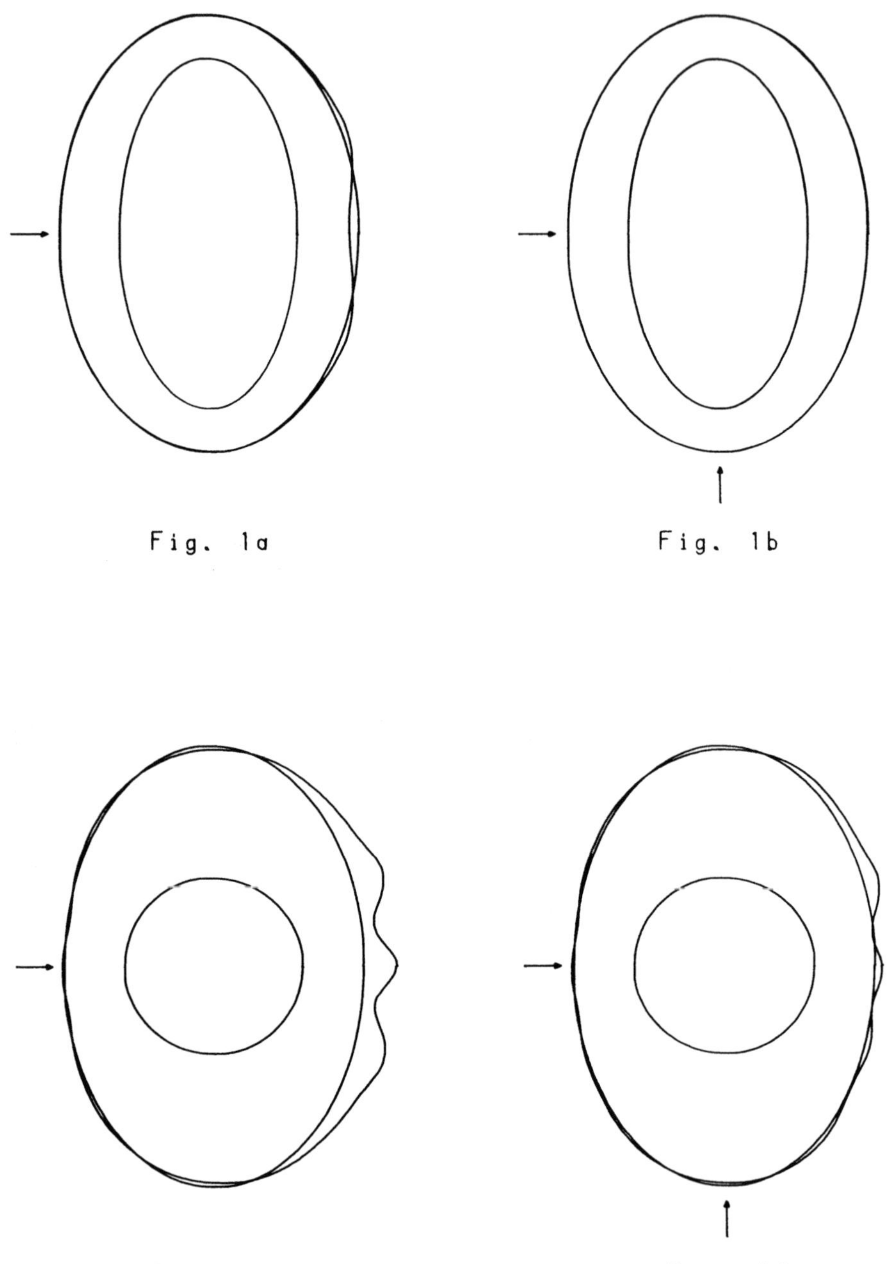

Fig. 1a

Fig. 1b

Fig. 1c

Fig. 1d

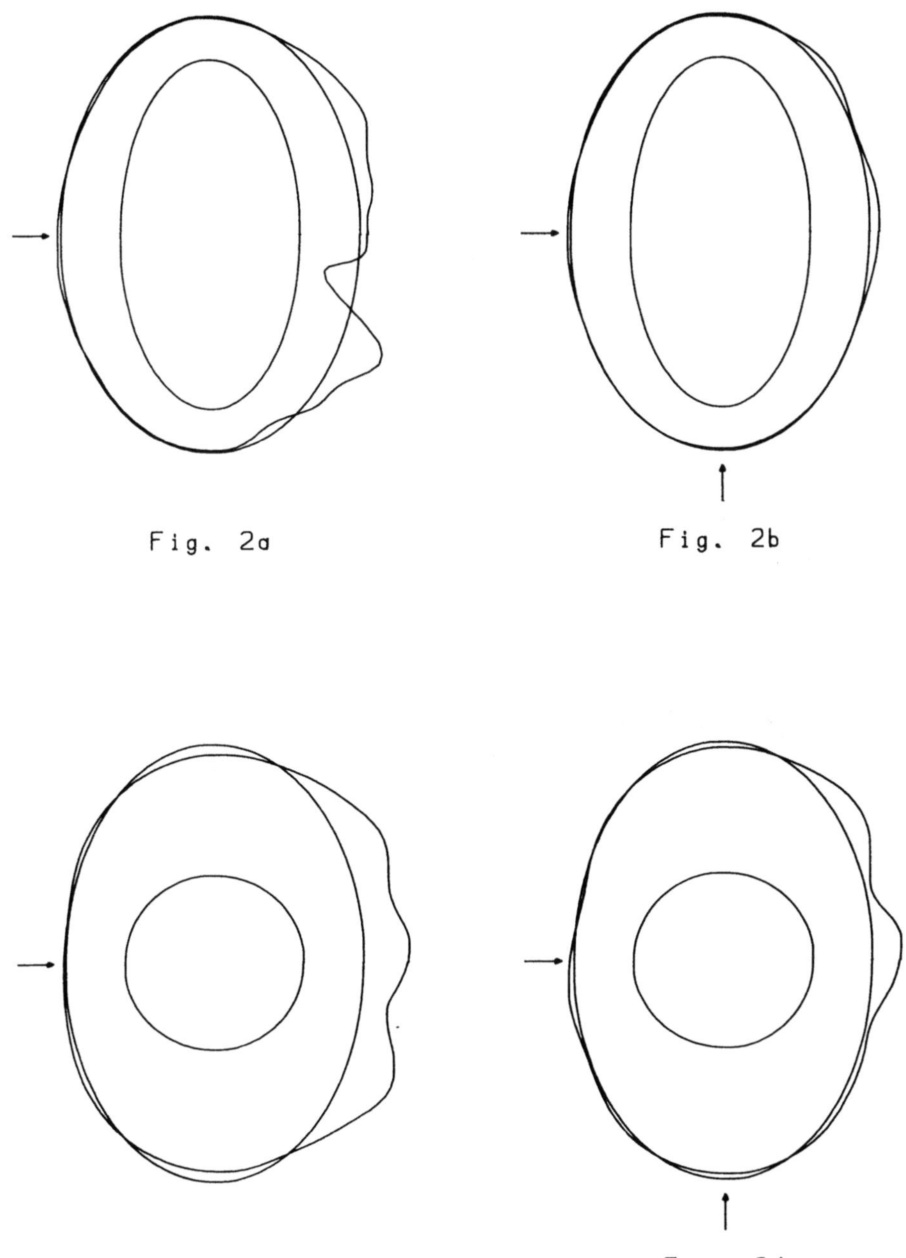

Fig. 2a　　　　Fig. 2b

Fig. 2c　　　　Fig. 2d

4. Colton, D. and Sleeman, B.: Uniquesness theorems for the inverse problem of acoustic scattering. IMA J. Applied Math. 31 (1983), 253–259.

5. Jones, D.S.: Note on a uniqueness theorem of Schiffer. Applicable Analysis 9 (1985), 181–188.

6. Kirsch, A.: Generalized boundary value and control problems for the Helmholtz equation, Habilitationsschrift, Universität Göttingen, 1984.

7. Kress, R. and Mohsen, A.: On the simulation source technique for exterior problems in acoustics. Math. Meth. in the Appl. Sci. (to appear).

8. Kussmaul, R.: Ein numerisches Verfahren zur Lösung des Neumannschen Aussen-raumproblems für die Helmholtzsche Schwingungsgleichung. Computing 4 (1969), 246–273.

9. Limic, N.: Galerkin–Petrov method for Helmholtz equation exterior problems. Glasnik Matematicki 36 (1981), 245–260.

10. Millar, R.F.: The Rayleigh hypothesis and a related least–squares solution to scattering problems for periodic surfaces and other scatterers. Radio Science 8 (1973), 785–796.

11. Müller, C.: Grundprobleme der mathematischen Theorie elektromagnetischer Schwingungen. Springer, Berlin 1957.

12. Müller, C. and Kersten, H.: Zwei Klassen vollständiger Funktionensysteme zur Behandlung von Randwertaufgaben der Schwingungsgleichung $\Delta U + k^2 U = 0$. Math. Meth. in the Appl. Sci. 2 (1980), 48–67.

13. Sleeman, B.D.: The inverse problem of acoustic scattering. IMA J. Applied Math. 29 (1982), 113–142.

Andreas Kirsch and Rainer Kress
Institut für Numerische und Angewandte Mathematik
Universität Göttingen
Lotzestr. 16–18
D–3400 Göttingen

International Series of
Numerical Mathematics, Vol. 77
© 1986 Birkhäuser Verlag Basel

APPROXIMATION OF DISCRETE
PROBABILITY DISTRIBUTIONS
IN SPHERICAL STEREOLOGY

Rudolf Gorenflo

To Professor Dr. F. Reutter on the occasion of his 75th birthday

Abstract. If in spherical stereology the actual
radius of spheres obeys a discrete probability
distribution with unknown jump points the solution of the relevant Abel integral equation is not
continuous, and hence the supremum norm is inap-
propriate for estimating the error of an approximation. We show that the backward Euler method
converges in the L^1-norm, also in the more general case that the distribution function is a
linear combination of Heaviside functions superimposed on a Lipschitz-continuous background. We
treat both cases:
(a) cutting plane (first kind integral equation),
(b) cutting slice (second kind integral equation).
For convenience (in order to avoid a superficial
nonlinearity) we work with number densities and
corresponding cumulative functions instead of with
probability densities and distribution functions.

1. THE INTEGRAL EQUATIONS

Spherical stereology is concerned with the determination
of the probability distribution of the radius of spheres randomly
embedded in a solid opaque medium. A plane cut is made through the
medium, and from the observed probability distribution of the radius ρ (called the "apparent radius") of circles cut out from
spheres hit by the cutting plane one wants to obtain information
on the probability distribution of the radius r (called the "actual radius") of spheres in the medium. This problem which seems
first to have been described by Wicksell in [16] is important in
very diverse fields (see [9], [13], [15]) and naturally has attrac-
ted the attention of numerical analysts (see [1]). The problem
leads to an Abel type integral equation for whose presentation we
follow Reid [14] who works with *number densities* instead of probab-

ility densities, thus avoiding a superficial nonlinearity inherent in Wicksell's seminal paper [16]. The transition from these number densities to probability densities can be achieved by multiplying with a scaling factor.

We introduce number densities $\hat{f}(r)$ and $\hat{g}(\rho)$, where (in understandable differential notation)

$\hat{f}(r)dr$ = mean number of spheres with radius in $[r,r+dr)$ and centers in a volume of size 1,

$\hat{g}(\rho)d\rho$ = mean number of circles with radius in $[\rho,\rho+d\rho)$ and centers in an area of size 1 of the cutting plane.

By methods of geometrical probability a *first kind Abel integral equation* is derived which allows to determine $\hat{f}(r)$ for $0 \leq r \leq \bar{r}$ if $\hat{g}(\rho)$ for $0 \leq \rho \leq \bar{r}$ is known:

$$(1.1) \qquad 2\rho \int_{r=\rho}^{\bar{r}} (r^2-\rho^2)^{-1/2} \; \hat{f}(r)dr = \hat{g}(\rho).$$

Here $\bar{r}$ is an upper bound to the possible actual radius r (hence also to the possible apparent radius ρ). In case $\bar{r}=\infty$ we should replace "$\leq \bar{r}$" by "$< \infty$"; our interest, however, will be the case $0 < \bar{r} < \infty$.

Bach [2] considered a variant leading, in contrast to (1.1), to a *second kind Abel integral equation*. He proposed to cut out from the solid medium a sclice of thickness $s > 0$ (lying between two parallel planes) so thin, however, that light can fall through producing shadows of the parts of spheres cut out. If the light is perpendicular to the slice and the shadows are thrown onto a plane of observation parallel to the slice we can again make statistics on the apparent radius of circles, $\hat{g}(\rho)$ dρ now being the mean number of circles with radius in $[\rho,\rho+d\rho)$ and centers in an area of size 1 of the plane of observation. Bach derived the integral equation

$$(1.2) \qquad s\hat{f}(\rho)+2\rho \int_{\rho}^{\bar{r}} (r^2-\rho^2)\hat{f}(r)dr = \hat{g}(\rho),$$

and called the problem of determining $\hat{f}(r)$ for $0 \leq r \leq \bar{r}$ from given $\hat{g}(\rho)$ for $0 \leq \rho \leq \bar{r}$ the "*tomato salad problem*".

105

Please note that (1.2) for s = 0 yields (1.1). We therefore shall treat, as long as feasible (1.2) and (1.1) simultaneously by just considering the integral equation (1.2) under the condition

(1.3) $s \geq 0$.

Furthermore, we shall assume

(1.4) $0 < \bar{r} < \infty$.

We refer the interested reader to Goldsmith [10] for a treatment of the tomato salad problem by an integral equation connecting the probability densities instead of the number densities.

Substituting (*area instead of radius*)

(1.5) $\begin{cases} a = r^2, \ \alpha = \rho^2, \ 0 \leq a \leq \bar{a} = \bar{r}^2, \ 0 \leq \alpha \leq \bar{a}, \\ \hat{f}(r)\,dr = f(a)\,da, \ \ \hat{g}(\rho)\,d\rho = g(\alpha)\,d\alpha \end{cases}$

in (1.2) we get an Abel integral equation with a simpler kernel:

(1.6) $sf(\alpha) + \displaystyle\int_{\alpha}^{\bar{a}} (a-\alpha)^{-1/2} f(a)\,da = g(\alpha)$.

The further substitutions

(1.7) $\begin{cases} t = \bar{a} - a, \ x = \bar{a} - \alpha, \ 0 \leq t \leq \bar{a}, \ 0 \leq x \leq \bar{a}, \\ f(\bar{a}-t) = f^*(t), \ g(\bar{a}-x) = g^*(x) \end{cases}$

then transform (1.6) into the integral equation

(1.8) $sf^*(x) + \displaystyle\int_{0}^{x} (x-t)^{-1/2} f^*(t)\,dt = g^*(x)$.

For subsequent analysis of discretization it will be convenient to work with the *cumulative functions* (for $0 \leq t \leq \bar{a}$, $0 \leq x \leq \bar{a}$)

(1.9) $\varphi(t) = \displaystyle\int_{0}^{t} f^*(\tilde{t})\,d\tilde{t}, \ \gamma(x) = \displaystyle\int_{0}^{x} g^*(\tilde{x})\,d\tilde{x}$

which are "smoother" then the densities f^* and g^*. By integrating (1.8) we obtain the integral equation (for $0 \leq x \leq \bar{a}$)

(1.10) $s\varphi(x) + \displaystyle\int_{0}^{x} (x-t)^{-1/2} \varphi(t)\,dt = \gamma(x)$

which henceforth will be the one subjected to discretization. With g^* we also know γ, can determine φ and by "differentiation" f^*, then backsubstitute to obtain $\hat{f}$.

2. THE PROBLEM

Let it *a priori* be known that only values r_j for $j \in J$ where J is a finite or denumerably infinite index set can be taken on by the actual radius r. With Dirac's delta-function notation the number density $\hat{f}$ then has the form

$$(2.1) \qquad \hat{f}(r) = \sum_{j \in J} c_j \, \delta(r-r_j)$$

in which we assume both the coefficients c_j and the possible radii r_j as unknown and to be determined from $\hat{g}(\rho)$, $0 \leq \rho \leq \bar{r}$, via equation (1.2) or one of the transformed equations (1.6), (1.8), (1.10). For convenience of analysis we assume

$$(2.2) \qquad 0 < r_j \leq \bar{r} \quad \text{for} \quad j \in J \ ,$$

and naturally we have

$$(2.3) \qquad c_j \geq 0 \quad \text{for} \quad j \in J, \quad \sum_{j \in J} c_j = C < \infty.$$

We shall, instead of (2.3), more weakly assume

$$(2.3') \qquad c_j \in \mathbb{R} \quad \text{for } j \in J, \quad \sum_{j \in J} |c_j| = C < \infty.$$

With $a_j = r_j^2$, $\xi_j = \bar{a} - a_j$ then

$$(2.4) \qquad f^*(t) = \sum_{j \in J} c_j \, \delta(t-\xi_j), \text{ all } 0 \leq \xi_j < \bar{a} \ ,$$

and with the Heaviside-function $H(y)$, $= 1$ for $y > 0$, $= 0$ for $y \leq 0$, we have

$$(2.5) \qquad \varphi(t) = \sum_{j \in J} c_j \, H(t-\xi_j) \ .$$

With $y_+ = \max(y,0)$ we find that γ is of the form

$$(2.6) \qquad \gamma(x) = s \sum_{j \in J} c_j \, H(x-\xi_j) + 2 \sum_{j \in J} c_j \, \sqrt{(x-\xi_j)_+}$$

Now think the data being collected by a *histogram technique*. By counting the number of apparent radii ρ between ρ' and $\rho'' > \rho'$ we obtain approximate values for

$$\int_{\rho'}^{\rho''} \hat{g}(\rho)\,d\rho,$$ hence with $x' = \bar{a}-\rho'^2$, $x'' = \bar{a}-\rho''^2$ for

$$\gamma(x')-\gamma(x'') = \int_{x''}^{x'} g^*(x)\,dx\,.$$ We thus can assume that the values of $\gamma(x)$ are approximately known for points x_j with $j = 1,2,\ldots,N$. For convenience, we shall assume these points to be equidistant, $x_j = jh$, $h = \bar{a}/N$; note that $\gamma(0) = 0$ by definition (1.9).

We now state the problem to be treated.

Problem: *For $N \in \mathbb{N}$, $h = \bar{a}/N$, $x_j = jh$, and given values $\gamma(x_j)$ for $j = 1,2,\ldots,N$, find an approximation φ_h to the solution φ of the integral equation (1.10) for $s \geq 0$ under the assumption that φ is of the form (2.5) and (2.3') is fulfilled. Investigate for convergence as $N \to \infty$. Analyze amplification of noise in case of inaccurately given values $\gamma(x_j)$.*

Convention: *We assume (in points of discontinuity) the functions φ and γ (the latter in the case $s > 0$) to be continuous from the left.*

3. DISCRETIZATION

Let V be the *space of functions* $v : [0,\bar{a}] \to \mathbb{R}$, *of bounded variation, left-continuous and with* $v(0) = 0$. *With* $N \in \mathbb{N}$, $h = \bar{a}/N$, *grid-points* $x_j = jh$ *(for* $0 \leq j \leq N$, $j \in \mathbb{N}$*) we define five operators*, namely an <u>integral</u> <u>operator</u> $A : V \to V$, a <u>restriction operator</u> $R_h : V \to \mathbb{R}^N$, an <u>extension</u> <u>operator</u> $E_h : \mathbb{R}^n \to V$, a <u>projection operator</u> $P_h : V \to V_h$, and an <u>approximate</u> <u>integral</u> <u>operator</u> $A_h : V \to V$. *By* V_h *we denote the space of step-functions* $x \to w(x)$ *with* $w(0) = 0$ *and* w *being constant in each subinterval* $(x_{j-1}, x_j]$ *for* $j = 1,2,\ldots,N$. *We have* $V_h \subset V$ *and* $\dim V_h = N$.

The operators are defined as follows.

$$(3.1) \qquad (Av)(x) = \int_0^x (x-t)^{-1/2} v(t)\,dt, \quad 0 \leq x \leq \bar{a}.$$

$$(3.2) \qquad R_h v = \begin{pmatrix} v(x_1) \\ v(x_2) \\ \vdots \\ v(x_N) \end{pmatrix}.$$

$$(3.3) \qquad (E_h \vec{\psi})(x) = \begin{cases} 0 & \text{for } x = 0 \\ \\ \psi_j & \text{for } x_{j-1} < x \le x_j \ . \end{cases}$$

By ψ_j the components of the vector $\vec{\psi}$ are denoted.

$$(3.4) \qquad (P_h v)(x) = \begin{cases} 0 & \text{for } x = 0 \\ \\ v(x_j) & \text{for } x_{j-1} < x \le x_j \end{cases}$$

$$(3.5) \qquad (A_h v)(x) = \int_0^x (x-t)^{-1/2} (P_h v)(t)\,dt, \quad 0 \le x \le \bar{a}$$

Observe that we have

$$(3.6) \qquad P_h = E_h R_h, \quad A_h = A P_h = A E_h R_h \ .$$

Lemma 3.1: With the *semi-circulant matrix*

$$(3.7) \qquad B_h = \begin{pmatrix} b_0 & & & & \\ b_1 & b_0 & & & \\ b_2 & b_1 & b_0 & & \\ \vdots & & & & \\ b_{N-1} & b_{N-2} & \cdots & b_1 & b_0 \end{pmatrix}$$

where $b_\lambda = \sqrt{\lambda+1} - \sqrt{\lambda}$ we have

$$(3.8) \qquad R_h A_h = 2\sqrt{h}\, B_h R_h \ .$$

Proof: The j-th component of $R_h A_h v$ is

$$(R_h A_h v)_j = \int_0^{x_j} (x_j - t)^{-1/2} (P_h v)(t)\,dt$$

$$= \sum_{k=1}^{j} v(x_k) \int_{x_{k-1}}^{x_k} (x_j - t)^{-1/2}\,dt = 2\sqrt{h}\, b_{j-k} \ .$$

Note that $b_0 = 1$ and hence B_h^{-1} exists and is semi-circulant, too.

With I and I_h as identity mappings in V and $\mathbb{R}^N$, respectively, we write (1.10) as

(3.9) $\qquad (sI+A)\varphi = \gamma$

and try to find $\varphi_h \in V_h$ such that

(3.10) $\qquad R_h(sI+A_h)\varphi_h = R_h\gamma.$

This leads, in view of Lemma 3.1, to the <u>approximation procedure</u>:

Determine $\vec{\psi}_h \in \mathbb{R}^N$ from

(3.10') $\qquad (sI_h+2\sqrt{h}\ B_h)\vec{\psi}_h = R_h\gamma$

and extend $\vec{\psi}_h$ to

(3.11) $\qquad \varphi_h = E_h\vec{\psi}_h.$

Because $sI_h+2\sqrt{h}\ B_h$ is semicirculant with positive diagonal entries $\vec{\psi}_h$ exists uniquely and can be calculated as solution of a triangular system of linear equations.

4. CONVERGENCE

Because the solution φ of (1.10) is expected to have jumps of unknown height at unknown points the *supremum-norm* (so often used by researchers in numerical solution techniques for Abel integral equations) is *not suitable*. See [11] and [12] for discussions on this topic. As already proposed in [12] we shall use the L^1-norm for analysis of convergence, here as follows.

Take

(4.1) $\qquad \| v \|_1 = \int_0^{\overline{a}} |v(x)|\ dx \quad \text{for } v \in V,$

(4.2) $\qquad \| \vec{\psi} \|_{1,h} = h \sum_{j=1}^N |\psi_j| \quad \text{for } \vec{\psi} \in \mathbb{R}^N.$

The factor h in the definition of $\| \cdot \|_{1,h}$ is used for reasons of compatibility.

We want to find an estimate for $\| \varphi - \varphi_h \|_1$. From $\| \varphi - \varphi_h \|_1 \leq \| \varphi - P_h\varphi \|_1 + \| P_h\varphi - \varphi_h \|_1$ we conclude

110

(4.3) $\qquad \|\varphi-\varphi_h\|_1 \leq hT + \|R_h\varphi - \vec{\psi}_h\|_{1,h}$

with $\vec{\psi}_h$ as solution of (3.10') and T as total variation of φ.

Because of linearity and (2.3') we first consider the expression $\|R_h\varphi - \vec{\psi}_h\|_{1,h}$ for the *very special case*

(4.4) $\qquad \varphi(x) = H(x-\xi), \ 0 \leq \xi < \bar{a}.$

If we succeed in obtaining an estimate independent of ξ we simply have to multiply with $C = \sum\limits_{j \in J} |c_j|$ to have an estimate for the general case (2.5).

So assume (4.4) for the following calculations. Abbreviate

(4.5) $\qquad \vec{\varepsilon}_h = \vec{\psi}_h - R_h\varphi, \qquad \vec{e}_h = (sI_h + 2\sqrt{h}\ B_h)\ \vec{\varepsilon}_h.$

These are vectors in $\mathbb{R}^N$. From (see(3.10') and (3.9))

$$(sI_h + 2\sqrt{h}\ B_h)\ \vec{\psi}_h = R_h\gamma = R_h(sI + A)\varphi$$

and (use(3.8))

$$(sI_h + 2\sqrt{h}\ B_h)R_h\varphi = (s\ R_h + R_hA_h)\varphi = R_h(sI+A_h)\varphi$$

we deduce (remember (3.6): $A_h = A\ P_h$)

(4.6) $\qquad \vec{e}_h = R_h(A - A_h)\varphi = R_hA(\varphi - P_h\varphi).$

Now for ξ fixed in $(0,\bar{a})$ there exists an index $m \in \{1,2,\ldots,N\}$ such that $x_{m-1} \leq \xi < x_m$. For our very special function (4.4) we have

$$|(\varphi - P_h\varphi)(t)| \leq \begin{cases} 0 \text{ for } t \notin (x_{m-1},x_m] \\ 1 \text{ for } t \in (x_{m-1},x_m] \end{cases},$$

and for the j-th component

$$(e_h)_j = \int_0^{x_j} (x_j-t)^{-1/2}(\varphi(t) - P_h\varphi(t))dt$$

of the vector $\vec{e}_h$ we get

$$(e_h)_j = 0 \quad \text{for} \quad j \le m-1 \; ,$$

$$|(e_h)_j| \le \int_{x_{m-1}}^{x_m} (x_j-t)^{-1/2} dt = 2\sqrt{h} \; b_{j-m} \quad \text{for} \quad j \ge m \; .$$

It follows by Lemma 3.1 that

$$\|\vec{e}_h\|_{1,h} \le h \cdot 2\sqrt{h} \sum_{j=m}^{N} b_{j-m} \le 2h^{3/2} N^{1/2} \; ,$$

$$(4.7) \qquad \|\vec{e}_h\|_{1,h} \le 2\sqrt{a}\, h \; ,$$

and (4.5) now yields

$$(4.8) \qquad \|\vec{\varepsilon}_h\|_{1,h} \le \|(s\, I_h + 2\sqrt{h}\, B_h)^{-1}\|_{1,h} \|\vec{e}_h\|_{1,h} \; ,$$

the first norm on the right hand side meaning the 1-matrix norm
for $N \times N$ matrices.

At this instant we must *distinguish first and second kind*
Abel integral equations.

<u>Case</u> (i): $s = 0$, <u>Case</u> (ii) $s > 0$.

Eggermont has shown in [6] (see also [7] and compare [5])
by aid of results of results of [4] and [8] that $\|B_h^{-1}\|_\infty$ is
uniformly bounded (independant of h), whereas Bakke and Jackiewicz
in [3] (their case $\Theta = 1$) have shown that $\|(s\, I_h + 2\sqrt{h}\, B_h)^{-1}\|_\infty$
is uniformly bounded (independent of h) for <u>fixed</u> $s > 0$. Now happily
for semi-circulant matrices the 1-norm coincides with the ∞-norm,
and we can use what these authors have proved for our purposes.
From (4.7) and (4.8) we can infer the existence of a constant
$M(s)$, depending solely on s but not on ξ, such that

$$(4.9 \; i) \qquad \|\vec{\varepsilon}_h\|_{1,h} \le \sqrt{a}\, M(0) h^{-1/2} \quad \text{in case (i)} \; ,$$

$$(4.9 \; ii) \quad \|\vec{\varepsilon}_h\|_{1,h} \le \sqrt{a}\, M(s) h \quad \text{in case (ii)} \; .$$

Freeing ourselves from the very special function (4.4)
and taking (4.3) into account we can state a theorem.

<u>Theorem 4.1</u>: *Assume that the integral equation* (1.10)
with fixed $s \ge 0$ *has a solution* φ *of the form* (2.5) *with all*
$\xi_j \in [0,\bar{a})$ *and that* (2.3') *is fulfilled. Then the approximation*

procedure described in section 3. (which may be called EULER BACKWARDS or EULER IMPLICIT) yields an approximate solution φ_h, and for $h \to 0$ we have (in Landau's order notation)

$$\| \varphi - \varphi_h \|_1 = \begin{cases} O(\sqrt{h}) & \text{if } s = 0 \\ O(h) & \text{if } s > 0 . \end{cases}$$

Corollary 4.1: *If instead of φ_h a function $\tilde{\varphi}_h$ is calculated from an erroneous data function $\tilde{\gamma}$ we have*

$$\| \varphi - \tilde{\varphi}_h \|_1 = \begin{cases} O(\sqrt{h}) + O(\frac{1}{\sqrt{h}}) \, \| R_h(\gamma - \tilde{\gamma}) \|_{1,h} & \text{if } s = 0 \\ O(h) + O(1) \, \| R_h(\gamma - \tilde{\gamma}) \|_{1,h} & \text{if } s > 0 . \end{cases}$$

The <u>proof</u> of the corollary is left to the reader.

<u>5</u>. MIXED CUMULATIVE FUNCTION

In stereology it may happen that the number density of the actual radius of spheres is a *linear combination of delta functions superimposed on a piecewise continuous background:*

$$(5.1) \qquad \hat{f}(r) = \sum_{j \in J} c_j \, \delta(r - r_j) + \hat{f}_0(r) ,$$

$\hat{f}_0$ piecewise continuous and $\Sigma |c_j| = C < \infty$. Going through the transformations of section <u>1</u>. we find that the function φ_0 corresponding to $\hat{f}_0$ is *Lipschitz-continuous* if furthermore $\hat{f}(r) = O(r)$ as $r \to 0$. We can apply the approximation procedure of section <u>3</u>.

For analysis of convergence because of linearity it now suffices to investigate the case $\hat{f} = \hat{f}_0$ which corresponds to $\varphi = \varphi_0$.

Assuming φ as *Lipschitz-continuous* with Lipschitz constant L we imitate the analysis of section <u>4</u>. We have

$$| (e_h)_j | \leq \sum_{k=1}^{j} \int_{x_{k-1}}^{x_k} (x_j - t)^{-1/2} \, | \varphi(t) - P_h \varphi(t) | \, dt$$

$$\leq Lh \sum_{k=1}^{j} b_{j-k} \, 2 \sqrt{h} = 2 L h^{3/2} \sqrt{j} ,$$

$$\| \vec{e}_h \|_{1,h} \leq 2\ Lh^{5/2} \sum_{j=1}^{N} \sqrt{j} \leq 2\ Lh^{5/2}\ N^{3/2}$$

$$= 2\ L\ \bar{a}^{3/2}\ h = O(h)$$

as in (4.7). We thus obtain the same asymptotic statement as in Theorem 4.1.

Theorem 5.1: *In Theorem 4.1 the assumption on the structure of the solution φ can be replaced by the assumption*

$$(5.2) \quad \varphi(t) = \sum_{j \in J} c_j H(t-\xi_j) + \varphi_o(t),\quad 0 \leq t \leq \bar{a},\quad \sum_{j \in J} |c_j| < \infty, \text{all } \xi_j \in [0,\bar{a}).$$

where φ_o is Lipschitz continuous.

6. A NUMERICAL CASE STUDY

On a Texas Instrument programmable pocket calculator TI 59 for $s=0$ and $s=1$ the equation (1.10) was treated with steplength $h=1/10$ (that is $N=10$). The right hand side was $\gamma(x) = sH(x-\xi) + 2\sqrt{(x-\xi)_+}$, $0 \leq x \leq \bar{a} = 1$, with $\xi = 0.57142\ 85714$. The following table of results was obtained (in Oberwolfach on 19th May 1986 *between lunch and coffee*). Note that the exact solution is $\varphi(x) = H(x-\xi)$.

$j=x/h$	$\varphi(x)$	$\varphi_h(x)$ for $s=0$	$\varphi_h(x)$ for $s=1$
$1 \leq j \leq 5$	0	0	0
6	1	0.53452	0.81966
7	1	0.91249	0.92034
8	1	0.96400	0.94968
9	1	0.98010	0.96402
10	1	0.98714	0.97247

REFERENCES

[1] R.S. Anderssen and A.J. Jakeman: Product integration for par-
     ticle size distributions. Utilitas Mathematica $\underline{8}$ (1975),
     111-126.

[2] G. Bach: Über die Größenverteilung von Kugelschnitten in
     durchsichtigen Schnitten endlicher Dicke. Zeitschrift für
     wissenschaftliche Mikroskopie $\underline{64}$ (1959), 265-270.

[3] V.L. Bakke and Z. Jackiewicz: Stability analysis of product
     Θ-methods for Abel integral equations of the second kind.
     Num. Math. $\underline{48}$ (1986), 127-136.

[4] N.G. de Bruijn and P. Erdös: On a recursion formula and on
     some Tauberian theorems. Journal of Research of the National
     Bureau of Standards $\underline{50}$ (1953), 161-164.

[5] R.F. Cameron and S. McKee: The analysis of product integra-
     tion methods for Abel's equation using discrete fractional
     differentiation. IMA Journal of Numerical Analysis $\underline{5}$ (1985),
     339-353.

[6] P.P.B. Eggermont: A new analysis of the Euler-, midpoint-
     and trapezoidal- discretization methods for the numerical
     solution of Abel-type integral equations. Medical Image Pro-
     cessing Group Technical Report No. MIPG34 September 1979.
     State University of New York at Buffalo, Department of Com-
     puter Science.

[7] P.P.B. Eggermont: A new analysis of the trapezoidal- discre-
     tization method for the numerical solution of Abel-type in-
     tegral equations. Journal of Integral Equations $\underline{3}$ (1981),
     317-332.

[8] P. Erdös, W. Feller, and H. Pollard: A property of power
     series with positive coefficients. Bulletin of the American
     Mathematical Society $\underline{55}$ (1949), 201-204.

[9] R.L. Fullman: Measurement of particle sizes in opaque bodies.
     Journal of Metals $\underline{5}$ (1953), 447-452.

[10] P.L. Goldsmith: The calculation of true particle size dis-
     tributions from the sizes observed in a thin slice. British
     Journal of Applied Physics $\underline{18}$ (1967), 813-830.

[11] R. Gorenflo: How to compute very rough solutions of Abel in-
     tegral equations? International Series of Numerical Mathe-
     matics $\underline{73}$ (1985), 281-282.

[12] R. Gorenflo: Abel integral equations: applications-motivated
     solution concepts. To appear in Methoden und Verfahren
     der Mathematischen Physik.

[13] W.C. Krumbein: Thin section mechanical analysis of inducted
sediments. Journal of Geology <u>43</u> (1935), 482-496.

[14] W.P.Reid: Distribution of sizes of spheres in a solid from
a study of slices of the solid. Journal of Mathematics and
Physics <u>34</u> (1955), 95-102.

[15] E. Scheil: Die Berechnung der Anzahl und Größenverteilung ku-
gelförmiger Kristalle in undurchsichtigen Körpern mit Hilfe
der durch einen ebenen Schnitt erhaltenen Schnittkreise.
Zeitschrift für anorganische und allgemeine Chemie <u>201</u> (1931),
259-264.

[16] S.D. Wicksell: The corpuscle problem. A mathematical study
of a biometric problem. Biometrika <u>17</u> (1925), 84-99.

Rudolf Gorenflo
Fachbereich Mathematik
Freie Universität Berlin
Arnimallee 2-6
D-1000 Berlin 33
Germany

The author is member of the research group "Modelling and
Discretization" which is supported by the Kommission für Forschung
und wissenschaftlichen Nachwuchs of Free University of Berlin.

International Series of
Numerical Mathematics, Vol. 77

PARAMETER ESTIMATION FOR DISTRIBUTED SYSTEMS
ARISING IN FLUID FLOW PROBLEMS
VIA TIME SERIES METHODS

Tao Lin and Richard Ewing

Abstract. The process of determining unknown parameter values, such as porosity and permeability, which are necessary for computer models for fluid flow processes is very complex. The difficulties arising from multiphase flow problems are even more severe. There are many different sources of error in the modeling process which lead to the distributed parameter systems and their computer solution. Use of time series methods addresses several of these sources of error, allowing efficient procedures for large-scale time dependent problems which can often reduce the difficulty of the enormous nonlinear optimization problems arising from many parameter estimation techniques.

I. Introduction

The mathematical models which describe fluid flow processes contain function parameters which describe properties of the fluids or properties of the medium through which they flow together with interactions of the fluids and the medium. For more complex multiphase or multicomponent flows of fluids these function parameters can be highly non-linear, depending strongly upon the unknown functions in the partial differential equations (PDE's). Often these parameters cannot be measured directly, but must be determined indirectly from measurements of the pressure or flow rate of the fluids in a flow experiment.

The process of determining unknown media parameter values, such as porosity and permeability, which are used in a mathematical model for fluid flow to give a best fit to measured well production history is commonly called "history matching". This process leads to extremely large and ill-conditioned nonlinear optimization problems which are well known to be very difficult to solve. Instead of considering techniques to regularize and solve such complex optimization problems directly, we will focus upon attempts to simplify these problems in special cases when possible, and solve the simpler problems more accurately and efficiently.

Parameter identification problems have been discussed extensively in the literature (e.g., see [2,4,5,6]). However, in the practice of computer simulation of real distributed systems, several observations should be made:

(a) It is very difficult to estimate function parameters directly in the PDE model. On the other hand, for computer simulation we do not need the complete function parameter, we only need the values of the parameters at grid points.

118

(b) In practice, the available data from a real distributed system are usually time-dependent. Hence, we need an efficient way to take advantage of the upcoming new data to modify and improve the estimates for the unknown parameters, and to make our computer model more reliable.

(c) The procedure of parameter identification usually leads to a very difficult non-linear optimization problem [3,4,5], in which the functional evaluation is very time consuming, and the number of unknowns in the optimization problem is usually very large.

Because of (a), we shall discuss how to estimate the parameters in the computer model instead of those in the PDE. Based on (b) and (c), we use time series and zoning methods to discuss the parameter identification problem for the computer model.

By estimating the parameters in a discrete computer model instead of those in the continuous PDE model, we are clearly making a type of discretization error in the estimate by allowing only a finite parameter specification. In Section II, we choose this parameter representation as if we had constant media properties on each grid block or point. If this technique were used in a large-scale three dimensional reservoir simulator such as those commonly in use in the petroleum industry, an ill-conditioned nonlinear least squares problem with 50,000 or more unknowns could result. While minimizing the modeling error by using a large parameter space, this technique generates an extremely difficult optimization problem. Therefore, the feature of the history-matching problem in reservoir simulation that distinguishes it from parameter identification problems in other fields of science and engineering is the large dimensionality of both the system state and the unknown parameter space. To reduce both the statistical uncertainty and the computational complexity, one must decrease the number of unknowns and, if possible, utilize any additional available information.

To reduce the number of variables and obtain a tractable problem, several effective schemes are being studied. One process called "zoning or zonation" divides the reservoir into smaller sets of zones, each of which is assumed to have reservoir properties which can be described in a functional manner with very few parameters in each zone. As the number of zones is decreased for computational tractability, the modeling error increases. The specification of an optimal level of zoning has been discussed by several authors (e.g., see [5]). One zoning technique is described in Section III of this paper.

An alternate way of reducing the size of the optimization problem is to try to break the problem up into smaller pieces and to design special flow experiments to isolate certain parameters and increase the sensitivity of the output to these parameters to obtain better estimates of the error. These ideas are crucial for multiphase flow problems where nonlinear relative permeability functions, capillary pressure curves exhibiting hysteresis effects, and dispersion coefficients must be determined. Laboratory experiments with reservoir fluids on reservoir cores can be very useful in isolating and accurately approximating these nonlinear parameters. These estimates can then be used directly in the field-scale history matches to estimate global properties like porosity and permeability.

One must also use as much prior geological information and geostatistics as possible in obtaining good initial guesses for the nonlinear optimization problems. Otherwise the ill-conditioning of the large nonlinear least-squares problems can produce many local minima of the output least-squares problem which cause severe difficulties for most search techniques. Bayesian techniques and "kriging" from geostatistics have also proved to be good methods for obtaining better initial guesses for the optimization problems. A brief survey of some of these methods was presented in [5].

II. Time Series Method

In order to illustrate the concepts of the time series and zoning methods that can be applied to complex reservoir simulation properties, we will discuss a simple one-dimensional parabolic PDE which can be used to model many fluid or heat flow problems. The extensions of the ideas presented to more complex problems should be apparent to the reader; the use of a simple model problem is mainly for expositional clarity.

The distributed system discussed here is described by the following PDE model:

$$(1) \qquad \begin{cases} \text{PDE}: & u_t(x,t) = \Big(a(x)u_x(x,t)\Big)_x \,, & 0 < x < 1 \,,\ t > 0 \,, \\ \text{IC}: & u(x,0) = \varphi(x) \,, & 0 < x < 1 \,, \\ \text{BC}: & u(0,t) = E(t) \,,\ u(1,t) = 0 \,, & t > 0 \,. \end{cases}$$

The computer model we choose for this PDE is given by an explicit time-stepping scheme as follows:

$$(2) \qquad \mathbf{u}^{k+1} = \Phi\Big(a(x)\Big)\mathbf{u}^k + \begin{pmatrix} \frac{\tau}{h^2} A_1 E(k) \\ 0 \\ \vdots \\ 0 \end{pmatrix}$$

where we have the following definitions: h and τ are typical discretization lengths in space and time, respectively, and

$$(3) \qquad \begin{cases} \mathbf{u}^k & = & (u_1^k, \cdots, u_{N-1}^k)^T \\ u_j^k & \doteq & u(x_j, t_k) \\ E(k) & = & E(t_k) \,, \end{cases}$$

$$(4) \qquad \begin{cases} x_j & = & jh \,, & j = 0, \cdots, N \,, & Nh = 1 \\ t_k & = & k\tau \,, & k = 0, 1, \cdots \\ x_{j+\frac{1}{2}} & = & x_j + \frac{h}{2} \,, \end{cases}$$

$$(5) \quad \Phi\big(a(x)\big)$$

$$= \begin{pmatrix} 1 - \frac{\tau}{h^2}(A_1 + A_2) & \frac{\tau}{h^2}A_2 & & & \\ \frac{\tau}{h^2}A_2 & 1 - \frac{\tau}{h^2}(A_2 + A_3) & \frac{\tau}{h^2}A_3 & & \\ & \ddots & \ddots & \ddots & \\ & & & & \frac{\tau}{h^2}A_{N-1} \\ & & & \frac{\tau}{h^2}A_{N-1} & 1 - \frac{\tau}{h^2}(A_{N-1} + A_N) \end{pmatrix},$$

$$\text{where} \quad A_j = a(x_{j-\frac{1}{2}}) .$$

The parameter identification problem is proposed as follows:

<u>PIP</u>. Given

$$(6) \qquad\qquad u_x(0,t) = G(t) + \epsilon(t) ,$$

determine $\Phi(a(x))$ in (2), where $\epsilon(t)$ is some kind of measurement error.

Without loss of generality, we suppose

$$(7) \qquad a(x) \in \mathcal{A} = \{a(x) \in C(0,1)|0 < a_0 \leq a(x) \leq a_1 < \infty\} .$$

Then it is well known that we can find a constant C such that, for all $a(x)$ in $\mathcal{A}$, the solution of (2) approximates that of (1) within the error tolerance:

$$C(\tau + h^2) .$$

By discretizing (1) along the left boundary, we obtain

$$(8) \qquad u_x(0, t_{k+1}) \doteq \left(\frac{A_1}{ha(x_0)}, 0, \cdots, 0\right) u^{k+1} -$$
$$- \frac{A_1}{ha(x_0)} E(k+1) - \frac{h}{2\tau a(x_0)}[E(k+1) - E(k)] .$$

Using (6) and letting

$$(9) \qquad \begin{cases} B\big(a(x)\big) & = & \begin{pmatrix} \frac{\tau A_1}{h^2} \\ 0 \\ \vdots \\ 0 \end{pmatrix} \\[2em] H\big(a(x)\big) & = & \left(\frac{A_1}{ha(x_0)}, 0, \cdots, 0\right) \\ G(k) & = & G(t_k) \\ v(k+1) & = & -\frac{A_1}{ha(x_0)} E(k+1) - \frac{h}{2\tau a(x_0)}[E(k+1) - E(k)] , \end{cases}$$

we then have

$$(10) \qquad \begin{cases} \mathbf{u}^{k+1} & = & \Phi\big(a(x)\big)\mathbf{u}^k + B\big(a(x)\big)E(k) \\ G(k+1) & = & H\big(a(x)\big)\mathbf{u}^{k+1} + v(k+1) . \end{cases}$$

If we think of the $E(k)$'s and $G(k)$'s as the input and output, respectively, the system given by (10) is a single input and single output system. The following theorem gives the relationship between the input and output:

Theorem 1. *If $a(x)$ belongs to $\mathcal{A}$ given by (7), then system (10) is observable, and*

$$(11) \qquad G(k+N-1) + \alpha_{N-2}G(k+N-2) + \cdots + \alpha_0 G(k)$$

$$= v(k+N-1) + \alpha_{N-2}v(k+N-2) + \cdots + \alpha_0 v(k)$$

$$+ d_1 E(k+N-2) + \cdots + d_{N-1}E(k)$$

where

$$(12) \qquad |\lambda I - \Phi\big(a(x)\big)| = \lambda^{N-1} + \alpha_{N-2}\lambda^{N-2} + \cdots + \alpha_1\lambda + \alpha_0 \ ,$$

$$(13) \qquad \begin{pmatrix} d_1 \\ \vdots \\ \vdots \\ d_{N-1} \end{pmatrix} = \begin{pmatrix} 1 & 0 & \cdots & \cdots & 0 \\ \alpha_{N-2} & 1 & \cdots & \cdots & 0 \\ \alpha_{N-3} & \alpha_{N-2} & 1 & \cdots & 0 \\ \vdots & \vdots & \vdots & \cdots & \vdots \\ \alpha_1 & \alpha_2 & \cdots & \cdots & 1 \end{pmatrix} \Phi_* B\big(a(x)\big) \ ,$$

and

$$(14) \qquad \Phi_* \Phi\big(a(x)\big) \Phi_*^{-1} = \begin{pmatrix} 0 & 1 & 0 & \cdots & 0 \\ 0 & 0 & 1 & \cdots & 0 \\ \vdots & \vdots & \vdots & & \vdots \\ 0 & 0 & 0 & \cdots & 1 \\ -\alpha_0 & -\alpha_1 & -\alpha_2 & \cdots & -\alpha_{N-2} \end{pmatrix} \ .$$

Since $\{G(k)\}$, $\{E(k)\}$, $\{v(k)\}$ are known, we can apply a recursive least squares method [1] to estimate the α_i's and d_i's. Hence we can reduce <u>PIP</u> into one of the following inverse matrix eigenvalue problems:

<u>IEP 1</u>. Given

$$(15) \qquad f(\lambda) = \lambda^{N-1} + \alpha_{N-2}\lambda^{N-2} + \cdots + \alpha_1\lambda + \alpha_0 \ ,$$

find a Φ of form (5) such that

$$(16) \qquad |\lambda I - \Phi| = f(\lambda)$$

for any λ.

<u>IEP 2</u>. Given $\hat{\lambda}_1, \cdots, \hat{\lambda}_{N-1}$, find a Φ of form (5) such that the eigenvalues of Φ are $\hat{\lambda}_1, \cdots, \hat{\lambda}_{N-1}$.

Now, for any Φ of form (5), we make the following hypothesis:

$$(H1) \qquad A_1 \text{ of } \Phi \text{ is given and } A_i > 0 \ , \quad i = 2, \cdots, N \ .$$

Then we introduce a function:

$$(17) \qquad F(\Phi) = F(A_2, \cdots, A_N) = \Big(\lambda_1(\Phi), \cdots, \lambda_{N-1}(\Phi) \Big)^T$$

where $\lambda_i(\Phi)$'s are the eigenvalues of Φ. We notice that Φ has the following properties:

    1) Φ is symmetric,

    2) Φ is tridiagonal,

    3) all the subdiagonals of Φ are greater than a positive number $a_0 > 0$.

By these properties of Φ, we can easily prove:

Theorem 2. *The Jacobian*

$$(18) \qquad J\Big(F(\Phi)\Big) = \left(\frac{\partial \lambda_i}{\partial A_j} \right)$$

is always regular for any $(A_2, \cdots, A_N)^T > 0$. *Hence, if we apply Newton's method to solve* IEP 2, *we can obtain local quadratic convergence.*

Up to now, we have outlined the following procedure for identifying $\Phi(a(x))$ from the measurement $G(t)$ of $u_x(0, t)$:

STEP 1. Estimate α_i's from the given $G(t)$, $E(t)$ via the time series model given by Equation (11).

STEP 2. Estimate $\Phi(a(x))$ by solving IEP 1 or IEP 2.

The following theorem gives the continuous dependence of the estimated $\Phi(a(x))$ upon the given data.

Theorem 3. *For fixed data length M, the estimated $\Phi(a(x))$ by* STEP 1 *and* STEP 2 *is continuous with respect to the data*

$$\Big\{ G(k) \Big\}_0^M, \quad \Big\{ E(k) \Big\}_0^M, \quad \Big\{ v(k) \Big\}_0^M .$$

Usually, we do not know if IEP 2 has a solution or not. Then, in order to obtain an estimation of Φ, we can solve the following non-linear least squares problems instead of IEP 2:

$$(19) \qquad \min_{\Phi} \frac{1}{2} \sum_{i=1}^{N-1} \Big(\lambda_i(\Phi) - \hat{\lambda}_i \Big)^2$$

or

$$(20) \qquad \min_{\Phi} \frac{1}{2} \sum_{i=1}^{N-1} \Big(|\hat{\lambda}_i I - \Phi| - f(\hat{\lambda}_i) \Big)^2$$

where the $\hat{\lambda}_i$'s are given.

III. Zoning Method

For the parameter identification procedure developed in the previous sections, we notice two things: the good point is that it is very efficient to use the new data to modify the estimated parameters, but the bad point is that we have to solve a non-linear problem with as many unknowns as the number of grid points. The number of grid points required is determined by the error tolerance of the computer model, which we have chosen to be

$$C(\tau + h^2) \ .$$

If the tolerance is required to be less than $\frac{1}{1000}$, then h must be less than $\frac{1}{32}$. This means the number of grid points must be greater than 30, and the unknowns in IEP 1 and IEP 2 must be more than 30. It is difficult to solve IEP 1 or IEP 2 with more than 30 unknowns by Newton's method, even on large computers like the Cyber 760.

In order to apply the procedure in the last section, one idea is to choose a new computer model with less parameters. This leads us to apply a zoning technique in our computer model.

When we set up the computer model in the last section, we approximated $a(x)$ of (1) by a step function

$$(21) \qquad \tilde{a}(x) = A_i \ , \qquad x_{i-1} \le x < x_i \ .$$

This is not an extremely good approximation of $a(x)$.

Now, let us choose some new points evenly in $[0,1]$ as follows

$$(22) \qquad 0 = y_0 < y_1 < \cdots < y_{m-1} < y_m = 1 \ .$$

These points separate $[0,1]$ into m zones. In these zones, we want to approximate $a(x)$ by relatively smooth functions in such a way we can approximate $a(x)$ by a function $a^*(x)$ with less parameters than $\tilde{a}(x)$ given by (21). The obvious approach is to choose cubic spline functions on $[0,1]$. Using the methods of [7], we introduce four additional knots:

$$(23) \qquad \begin{cases} y_{-2} < y_{-1} < y_0 \\ y_m < y_{m+1} < y_{m+2} \ , \end{cases}$$

and let $h_1 = \frac{1}{m}$. Then as in [7], we can find $m + 3$ basis functions $\{B_i(x)\}_{i=-1}^{m+1}$ for S_3 $([0,1])$. Then the following lemma holds:

Lemma 4 [7]. *If $a(x) \in C^4 \, [0,1]$, then there exists*

$$(24) \qquad a^*(x) = \sum_{i=-1}^{m+1} b_i^* B_i(x)$$

such that

$$(25) \qquad \begin{cases} \|a - a^*\|_\infty & \leq \quad \frac{5}{384} \|a^{(4)}\|_\infty h_1^4 \\ \|a' - a^{*'}\|_\infty & \leq \quad \left[\frac{\sqrt{3}}{216} + \frac{1}{26}\right] \|a^{(4)}\|_\infty h_1^3 \,, \end{cases}$$

where

$$(26) \qquad \|g\|_\infty = \max_{1 \leq i \leq m} \; \sup_{x_{i-1} < x < x_i} |g(x)| \,.$$

If we use u_j^{*k} to approximate $u(x_j, t_k)$, we obtain a new computer model as follows:

$$(27) \qquad \begin{cases} \mathbf{u}^{*k+1} & = \quad \Phi\Big(a^*(x)\Big)\mathbf{u}^{*k} + B\Big(a^*(x)\Big)E(k) \\ G(k+1) & = \quad H\Big(a^*(x)\Big)\mathbf{u}^{k+1} + v^*(k+1) \\ \mathbf{u}^{*k} & = \quad \Big(u_1^{*k}, \cdots, u_{N-1}^{*k}\Big) \,, \end{cases}$$

where $\Phi(a^*(x))$, $B(a^*(x))$, $G(k)$, $v^*(k)$, $H(a^*(x))$ are defined in the same way as in previous sections by changing A_i into A_i^* and

$$(28) \qquad A_i^* = a^*(x_{j-\frac{1}{2}}) = \sum_{j=-1}^{m+1} b_j^* B_j(x_{j-\frac{1}{2}}) \,.$$

The following theorem tells us about the efficiency of the new computer model.

Theorem 5. *For $a(x) \in \mathcal{A}$, we can find a constant C such that*

$$(29) \qquad |u(x_i, t_j) - u_i^{*j}| \leq C(\tau + h^2) + C h_1^3 \,.$$

Hence, if

$$(30) \qquad h_1^3 \leq h^2 \,,$$

the new computer model is still within the error tolerance $C(\tau + h^2)$, so we can apply the procedure developed in the last section to estimate $\Phi(a^*(x))$ just by solving the following new non-linear problems.

<u>IEP 1*</u>. Given

$$f(\lambda) = \lambda^{N-1} + \alpha_{N-2}\lambda^{N-2} + \cdots + \alpha_1 \lambda + \alpha_0 \,,$$

find $b_{-1}, b_0, \cdots, b_{m+1}$ such that

$$|\lambda I - \Phi\big(a^*(x)\big)| = f(\lambda) \ .$$

An alternative problem is:

IEP 2*. Given $\hat{\lambda}_1, \cdots, \hat{\lambda}_{N-1}$, find $b_{-1}, b_0, \cdots, b_{m+1}$ such that the eigenvalues of $\Phi(a^*(x))$ are the $\hat{\lambda}_i$'s.

By the following theorem, we know to what extent we have reduced the number of unknowns in the non-linear problem:

Theorem 6. *If the error tolerance is*

$$C(\tau + h^2) = C(\tau + \frac{1}{N^2})$$

and

$$N^{1/3} \geq K \ ,$$

then we can choose m so that $m + 3$ (i.e., the number of unknowns in IEP 1* *or* IEP 2*) *is less than $\frac{N}{K} + 3$.*

Theorem 6 gives us a criterion to determine m in the new computer model. For a given error tolerance ϵ, we pick N to satisfy

$$\frac{1}{N^2} = h^2 < \epsilon \ ;$$

then choose the largest integer K such that

$$N^{1/3} > K$$

and let

$$m = \frac{N}{K} \ .$$

In this way, we obtain a comparison betweeen the number of unknowns in the standard procedure developed in the last section and in the zoning method:

Standard Method	Zoning Method	K
30	13	3
100	28	4
1000	103	10

From this table, we know that the more accuracy we require from the computer model, the more efficiency we obtain from the zoning method.

REFERENCES

[1] A. Albert, *Regression and the Moore-Penrose Pseudoinverse*, Academic Press (1972).

[2] H. T. Banks, A survey of some problems and recent results for parameter estimation and optimal control in delay and distributed parameter systems. *ICASE Report*, No. 81-26 (1981).

[3] L. Carotento and G. Raiconi, The finite element method in distributed parameter control systems. *Distributed Parameter Control Systems Theory and Application*, Pergamon Press (1982) 238-271.

[4] R. E. Ewing, Determination of Coefficients in Reservoir Simulation, <u>Numeri-cal Treatment of Inverse Problems for Differential and Integral Equations</u>, P. Deuflhardt and E. Hairer, eds., Birkhauser, Berlin (1982) 206-226.

[5] R. E. Ewing and J. H. George, Identification and control for distributed parameter in porous media flow. Distributed Parameter Systems. *Distributed Parameter Systems, Lecture Notes in Control and Information Sciences (M. Thoma, ed.)* Springer-Verlag, 70 (1985) 145-161.

[6] M. P. Polis, The distributed system parameter identification problem: A survey of recent results. *Third Symposium, Control of Distributed Parameter Systems* (1982).

[7] P. M. Prenter, *Splines and Variational Methods*, John Wiley and Sons (1975).

Tao Lin
Department of Mathematics
University of Wyoming
Laramie, WY 82071
U.S.A.

Richard Ewing
Departments of Mathematics
  and Petroleum Engineering
University of Wyoming
Laramie, WY 82071
U.S.A.

<u>Acknowledgement.</u> This research was supported in part by the U.S. Army Research Office Contract No. DAAG29-84-K002, by the U.S. Air Force Office of Scientific Research Contract No. AFOSR-85-0117, and by the National Science Foundation Grant No. DMS-8504360.

International Series of
Numerical Mathematics, Vol. 77
© 1986 Birkhäuser Verlag Basel

ANALYTIC SOLUTIONS TO THE INVERSE PROBLEM

OF THE NEWTONIAN POTENTIAL

Carlo Domenico Pagani [1]

Politecnico di Milano, p.za Leonardo da Vinci, 32, 20133 Milano (Italia)

ABSTRACT

Let G be a homogeneous body whose shape in R^3 is unknown. We want to determine the figure of G from measurements of the newtonian potential created by this body on a spherical surface surrounding G. We prove the existence of a (local) analytic solution to the stated problem.

INTRODUCTION

Let us consider a class of bodies G_u parametrized by smooth functions

$u: S^2 \to R$, $|u| < 1$, as follows (here is $S^2 = \{x \in R^2: x_1^2 + x_2^2 + x_3^2 = 1\}$). Let

$\phi_u: S^2 \to R^3$ be a differentiable embedding

$$S^2 \ni \omega \to \phi_u(\omega) = \omega \,(1+u(\omega))$$

Define $\Gamma_u = \phi_u(S^2)$ and G_u as the bounded domain whose boundary is Γ_u.

Let us denote by V_u the potential created by a mass of unitary density distributed over G_u (unknown):

$$V_u(x) = \int_{S^2} d\omega \int_0^{|\phi_u(\omega)|} |x-t\omega|^{-1} t^2 dt$$

Our problem consists in finding G_u from the knowledge of V_u on Γ_1 (the

sphere: $|x|=2$). Then, let us call B(u) the restriction of V_u on Γ_1 (pull-

___

[1] The present talk gives account of joint work with C. Maderna and S. Salsa [4] where one can find the proofs of the results stated here and more details.

128

back to S^2)

$$B(u) \equiv V_u \big|_{\Gamma_1}$$

We can formulate the problem as follows: <u>we are given</u> $v : S^2 \to R$, <u>and want to find</u> u <u>such that</u>

(1) $\qquad\qquad$ B(u) = v $\qquad$ <u>on</u> S^2

I know of no results on the existence of exact solutions to this problem. In some physical situations one has to determine an unknown surface, when the data are taken on the unknown surface itself: cfr., e.g., the Molodenski problem of physical geodesy, studied by Hörmander [1], a capacitor problem (D. Schaeffer, [7]), and the same inverse problem of the newtonian potential considered here, when the datum v is the restriction of V_u to Γ_u (cfr. [3]). All these cases have been dealt with by some version of the Nash-Moser technique, which is suitable for problems, like those mentioned above, involving "loss of derivatives". This technique can be characterized as an iterative method based on <u>global</u> linearization in contrast to the classical methods (such as the Banach contraction principle, the Newton-Kantorovich algorithm...) based on <u>local</u> linearization[2]. One crucial step in the application of the Nash-Moser scheme is the derivation of some estimates for the operator B, its first differential B', the inverse of the first differential, and for the non linear part of B (or the second differential of B): the so-called "tame estimates" (see [3]) which, together with the "smoothing operators", allow the convergence of the iterative procedure.

In the present problem, we still are faced with a loss of regularity when linearizing the equation (1); but a new difficulty arises, since the desired estimates for B' and its inverse cannot be derived in the usual sca-

---

[2] "Local" linearization means that the nonlinear equations is linearized about a vector running over a bounded subset, while "global" means that such subset is unbounded.

les of Hölder or Sobolev spaces; we are forced to work in Banach spaces of analytic functions. Now, a version of the Nash-Moser tecnique, suitable for analytic spaces, has been developed by Nirenberg [5]. Here, two families of Banach spaces X_s and Y_s (one containing the solutions, the other the data) are chosen depending on the parameter s ($0 \leq s \leq 1$) in such a way that $X_{s'} \subset X_s$ and $Y_{s'} \subset Y_s$ when $s' > s$, with continuous injection. The problem is to solve the equation $B(u) = v$, seeking u in some X_{s_∞} close to 0 when v is close to $B(0)$ (in our case, $B(0) = \pi/3$); here B is a map acting continuously from a ball in $X_{s'}$ into Y_s for every $s' > s$. To this purpose, starting at the point $u_o = 0$ one constructs in a suitable way a sequence of approximated solutions $u_k \in X_{s_k}$, with s_k tending from above to s_∞ in such a way that $B(u_{s_k})$ goes from $v_o = B(0)$ to v when k runs from 0 to $+ \infty$. Since the parameter s measures the degree of regularity of u_k, we see that each step in the iteration causes a loss of regularity: certain additional hypotheses on B, namely some bounds for the first differential of B, the inverse of the first differential, etc., allow us to keep under control this loss and make the sequence u_k to converge to the desired solution.

The application of this program to our situation meets the difficulty that we cannot choose two fixed scales of Banach spaces: we choose as X-spaces certain classes of analytic functions; however the definition of the Y-spaces (also classes of analytic fucntions, of course!) is part of the iteration procedure itself, in the sense that the approximate solution found at each step determines the Y-space for the next one. Then the crucial point is the estimate of the Y-norm at a given step in terms of the Y-norm at the preceding one. In Nirenberg's case this estimate is assured by the continuous injection of $Y_{s'}$ into Y_s, when $s'>s$. In our case this difficulty is overcome by means of some estimates concerning analytic solutions of the Dirichlet and Cauchy problems for a second order elliptic operator.

In this talk, I shall not enter into any detail of the procedure, but will limit myself to introduce the relevant function spaces, state the main result, and give few accounts on the linearized equation.

Let me conclude this lengthy introduction by recalling that another important question, that I do not touch here, is the question of the stability of the solutions. To this purpose, let me mention that, for the present problem, M.M. Lavrentev [2] proved that solutions with bounded gradient are logarithmically stable, while in [6] is proved that analytic solutions are (locally) hölder-stable.

STATEMENT OF THE MAIN RESULT

I first describe the relevant functional spaces. Let $\{S_j\}$ be a finite open covering of S^2 and $\{\psi_j\}$, $\psi_j: R^2 \supset D \to S_j$, a system of analytic local charts defined in the unit disc $D \equiv \{x \in R^2, |x| < 1\}$ and admitting a bounded analytic extension to the polydisc $D \times D \subset C^2$.

For s real, $0 \leq s \leq 1$ and $R > 0$ set

$$C^2 \supset \mathcal{D}_s^R \equiv \{ z = x + i\xi, \ |x| < 1, \ |\xi| < sR \}$$

Now, let f be a function defined on S^2. We say that $f \in X_s^R(S^2)$ if $f \circ \psi_j$ is analytic and admit a bounded analytic extension to $\mathcal{D}_s^R$ for all maps ψ_j. A norm is introduced by putting

$$||f||_s = \sup_j \ \sup_{z \in \mathcal{D}_s^R} \ |f \circ \psi_j(z)|$$

Let now f and u be in X_s (short notation for $X_s^R(S^2)$). We can solve uniquely the exterior Dirichlet problem

$$\Delta W = 0 \qquad \text{in } R^3 \setminus \bar{G}_u$$
$$W = f \qquad \text{on } \Gamma_u$$
$$W(x) \to 0 \qquad \text{as } |x| \to \infty$$

Call h the trace of W on Γ_1. The collection of all functions h defined in this way when f spans X_s forms a vector space hereby denoted $Y_{s,u}(S^2) \equiv Y_{s,u}$. We define a norm in $Y_{s,u}$ by putting

$$||h||_{s,u} \equiv ||f||_s$$

We can now state the main result.

THEOREM: <u>Fix s, 0<s<1; there exist two positive numbers R and</u> ε,ε <u>depending on s and R, such that, for every</u> $v \in Y_{1,0}$, <u>with</u> $||v||_{1,0} < \varepsilon$, <u>there is a unique solution</u> $u \in X_s^R(S^2)$ <u>of the equation</u> $B(u) = v$.

As I told, the proof of the theorem depends on a careful study of the first differential of the map $B(u)$. Then, take a smooth function $\rho: S^2 \to R$ and differentiate $B(u+t\rho)$ with respect to t, putting t=0 afterwards. Denoting the result by $B'(u)\rho$, we have

$$(2) \qquad B'(u)\rho(\omega) = \int_{S^2} |\phi_u(\omega')|^2 \rho(\omega') |2\omega - \phi_u(\omega')|^{-1} d\omega'$$

Then, the linearized equation $B'(u)\rho = h$ is a Fredholm integral equation of the first kind. Notice that $B'(u)\rho$ is the trace on Γ_1 of the single layer potential

$$W_u(x) = \int_{S^2} \rho(\omega') |\phi_u(\omega')|^2 |x - \phi_u(\omega')|^{-1} d\omega'$$

which is harmonic inside and outside G_u, continuous through Γ_u and its normal derivative suffers a jump across Γ_u of magnitude given by

$$(3) \qquad \rho(\omega) = \{ \partial_{N_u} W_u^+ \circ \phi_u(\omega) - \partial_{N_u} W_u^- \circ \phi_u(\omega) \} |D\phi_u(\omega)| |\phi_u(\omega)|^{-2}$$

where W_u^+ and W_u^- are the restrictions of W_u to the regions inside and outside G_u respectively, $|D\phi_u(\omega)|$ denotes the jacobian of the map ϕ_u, and N_u is the normal direction to Γ_u.

In such a way we get a representation of the invers of $B'(u)$ in terms of a solution of an elliptic problem.

In order to apply the iterative procedure, we need estimates for $B'(u) - B'(\tilde{u})$ (with u close to $\tilde{u}$) and estimates for the inverse of $B'(u)$; these estimates are actually derived from equations (2) and (3); they require the study of the Cauchy and Dirichlet problems for an elliptic operator.

REFERENCES

[1] L. Hörmander, The boundary problems of physical geodesy, Arch. Rat. Mech. Anal. 62 (1976) 1-52

[2] M.M. Lavrentev, Some inproperly posed problems of mathematical physics, Springer Tracts in Natural Phylosophy, 11, Springer Verlag, Berlin (1967)

[3] C. Maderna, C. Pagani and S. Salsa, Existence results in an inverse problem of potential theory, Nonlinear Anal. T.M.A., 10 (3) (1986) 277-298

[4] C. Maderna, C. Pagani and S. Salsa, Analytic solutions to the inverse problem of the newtonian potential, to appear on J. Math.Anal.and Appl.

[5] L. Nirenberg, An abstract form of the nonlinear Cauchy-Kowalewski theorem, J. Differential Geometry 6 (1972) 561-576

[6] C. Pagani, Stability of a surface determined from measures of potential, SIAM J. Math. Anal. 17 (1986)

[7] D.G. Schaeffer, the capacitor problem, Indiana Univ. Math. J. 24 (12) (1975) 1143-1167

International Series of
Numerical Mathematics, Vol. 77
© 1986 Birkhäuser Verlag Basel

A NOTE ON AN INVERSE PROBLEM
RELATED TO THE 3-D HEAT EQUATION

John R. Cannon and Salvador Pérez Esteva

Abstract. We consider the 3-D heat equation with over-specified data to obtain the heat source. For certain class of sources this problem is shown to be equivalent to the one-dimensional heat equation with an integral type condition. A continuous dependence result is proved.

1. INTRODUCTION

In [4] we considered the problem of finding $f = f(t)$ and $u = u(\bar{x},t)$ satisfying

$$
\text{I} \quad
\begin{cases}
u_t = \Delta u + f(t)\chi_D(\bar{x}), & \bar{x} \in \mathbb{R}^n, \quad t > 0 \\
u(\bar{x},0) = 0, & x \in \mathbb{R}^n, \\
u(\bar{0},t) = g(t), & t > 0,
\end{cases}
$$

where D is a fixed set in $\mathbb{R}^n$, χ_D is the characteristic function of D and function g is given. Existence and uniqueness were proved and some continuous dependence results were obtained for $n = 1,2$.

2. A PROBLEM IN 3-D

Let D be the ball of radius r_1 in $\mathbb{R}^3$, centered at the origin. Suppose (u,f) is the solution of (I). We have

$$
(2.1) \quad u(x,t) = \int_0^t \int_{\mathbb{R}^3} \frac{e^{-\frac{|\bar{x}-\bar{\xi}|^2}{4(t-\tau)}}}{(4(t-\tau))^{3/2}} \chi_D(\bar{\xi}) f(\tau) \, d\bar{\xi} \, d\tau
$$

Hence

$$
(2.2) \quad g(t) = u(\bar{0},t)
$$

$$
= \int_0^{r_1} r \int_0^t \frac{r\, e^{-\frac{r^2}{4(t-\tau)}}}{\sqrt{4\pi}\,(t-\tau)^{3/2}} f(\tau) \, d\tau \, dr
$$

$$134$$

We know that the function defined by

$$(2.3) \qquad v(x,t) = \frac{1}{\sqrt{4\pi}} \int_0^t \frac{x}{(t-\tau)^{3/2}} \, e^{-\frac{x^2}{4(t-\tau)}} f(\tau) \, d\tau$$

is the solution of

$$\text{II} \begin{cases} v_t = v_{xx} & x > 0, \quad t > 0 \\ v(x,0) = 0 & x > 0 \\ v(0,t) = f(t) & t > 0, \end{cases}$$

if f is continuous in $(0,\infty)$ and

$$|f(t)| \le Ct^{-\alpha} \quad t \in (0,\varepsilon), \quad 0 < \alpha < 1.$$

Then by (2.2) we have

$$\int_0^1 x \, v(x,t) \, dx = g(t) \qquad t > 0$$

Conversely suppose v satisfies

$$\text{III} \begin{cases} v_t = v_{xx} & x > 0, \quad t > 0 \\ v(x,0) = 0 & x > 0 \\ \int_0^1 x \, v(x,t) \, dx = g(t) & t > 0 \end{cases}$$

and assume v has the representation (2.3), then $v(0,t) = f(t)$ satisfies problem (I) with u as in (2.1).

LEMMA 2.1. **Let** v **be the solution of** (II) **and assume** v **can be written as in** (2.3), **then**

$$\lim_{x \to 0^+} x \, v_x(x,t) = 0$$

Proof: For $x > 0$ and $t > 0$ we have

$$(2.4) \qquad x \, v_x(x,t) = \frac{1}{\sqrt{4\pi}} \int_0^t \frac{x^3}{(t-\tau)^{3/2}} \, e^{-\frac{x^2}{4(t-\tau)}} f(\tau) \, d\tau$$

$$- \frac{1}{4\sqrt{\pi}} \int_o^t \frac{x^3}{(t-\tau)^{5/2}} \, e^{- \frac{x^2}{4(t-\tau)}} \, f(\tau) \, d\tau$$

$$= v(x,t) - \frac{1}{4\sqrt{\pi}} \int_o^t \frac{x^3}{(t-\tau)^{5/2}} \, e^{- \frac{x^2}{4(t-\tau)}} \, f(\tau) \, d\tau$$

Let

$$s(x,t) = \frac{1}{4\sqrt{\pi}} \frac{x^3}{t^{5/2}} \, e^{- \frac{x^2}{4t}}$$

and notice that

1) $s(x,t) > 0$ $x > 0, \quad t > 0$

2) $s(x,o+) = s(x,+\infty) = 0$

3)   For any $\delta > 0$

     $\lim\limits_{x \to o+} s(x,t) = 0$ uniformly on $t > \delta$

4) $\int_o^\infty s(x,t) \, dt = 1$

5)   For any $c > 0$

     $\lim\limits_{x \to 0+} \int_o^c s(x,t) \, dt = 1$

     It follows that

     $\lim\limits_{x \to o+} \int_o^t s(x,t-\tau) \, f(\tau) = f(t)$

and this proves the Lemma.

THEOREM 2.2.   Let $g = g(t)$ be continuously differentiable for $t > 0$ and $|g'(t)| \leq ct^{-\alpha}$ for $t \in (0,\varepsilon)$ and $0 < \alpha < 1$. Then there exists a unique solution $v = v(x,t)$ to problem III having the representation (2.3). The limit

$$f(t) = \lim_{x \to 0+} v(x,t)$$

is the solution of problem (I). Furthermore, for every $T > 0$, there exits $M > 0$ such that

$$\sup_{t \in (0,T]} |t^{\alpha} f(t)| \le M \sup_{t \in (0,T]} |t^{\alpha} g'(t)|$$

<u>Proof.</u> We look for a solution $v = v(x,t)$ of the form

$$(2.5) \quad v(x,t) = \frac{1}{\sqrt{4\pi}} \int_{0}^{t} \frac{x}{(t-\tau)^{3/2}} e^{-\frac{x^2}{4(t-\tau)}} f(\tau) \, d\tau$$

The condition in g and Lemma 2.1 imply that

$$(2.6) \quad g'(t) = \int_{0}^{r_1} x \, v_t(x,t) \, dx$$

$$= \int_{0}^{r_1} x \, v_{xx}(x,t) \, dx$$

$$= r_1 u_x(r_1,t) - u(r_1,t) + f(t)$$

By (2.4) and (2.6) we have

$$f(t) = g'(t) + \int_{0}^{t} s(r_1,t-\tau) f(\tau) \, d\tau,$$

with $s(x,t)$ as defined in Lemma 2.1.

Since $s(r_1,\cdot)$ is bounded, we see that

$$\left| \int_{0}^{t} s(r_1,t-\tau) f(\tau) d\tau \right| \le M \sup_{\tau \in (0,t)} |\tau^{\alpha} f(\tau)| \, t^{1-\alpha}$$

Hence for t small, the integral operator is a contraction and the theorem follows.

REFERENCES

1. J.R. Cannon, Determination of an Unknown heat source from overspecified data. SIAMJ. Numer. Anal., 5(1968).

2. J.R. Cannon, The One-Dimensional Heat Equation. Encyclopedia of Mathematics and its Applications, Vol. 29, 1984.

3. J.R. Cannon and R. E. Ewing, Determination of a source term in a linear parabolical partial differential equation. ZAMP, Vol. 27 (1976), 393-401.

4. J.R. Cannon, S. Pérez-Esteva, An inverse problem for the heat equation. J. Inverse Problems (Submitted).

5. G. Talenti, S. Vassella, A note on an ill-posed problem for the heat equation. J. Austral. Math. Soc. (series A) 32 (1982).

6. D. Widder, The Heat Equation. Academic Press, New York, 1975.

John R. Cannon
Washington State University
Pullman, WA 99163
U. S. A.

Salvador Pérez-Esteva
Instituto de Matemáticas
Universidad Nacional Autónoma de México
Circuito Exterior, Ciudad Universitaria
México 04510, D.F.
MEXICO

International Series of
Numerical Mathematics, Vol. 77
© 1986 Birkhäuser Verlag Basel

Undetermined Coefficient Problems for Nonlinear Elliptic and Parabolic Equations

Michael Pilant and *William Rundell*

Department of Mathematics

Texas A & M University

College Station, TX 77843.

Introduction.

In this talk we shall describe some numerical results arising from the problem of determining unknown coefficients that depend only on the function u in quasilinear second order partial differential equations of the form

$$-\Delta u = F(x, y, u, \nabla u, u_t) \tag{1.1}$$

defined in a certain region and subject to overposed boundary conditions on its boundary. We include the possibility of a term u_t in (1.1) since we intend to consider the case of parabolic as well as elliptic equations. This class of inverse problems has received a considerable amount of attention recently, and we refer the reader to references [1-7] for further details.

It will be convenient to consider the boundary data as belonging to two classes. The first consists of those conditions such that if the function F were known then there would be a well posed solution to (1.1) with these boundary conditions . We shall refer to these boundary and/or initial conditions as the primary data. For a given function F, the solution of (1.1) with the primary data will be referred to as the direct problem for u. The second class of boundary conditions consists of the overposed boundary data, and it is these conditions that we must use in order to recover certain unknown terms in the function F.

We shall first consider two special forms of the above equation in more detail. corresponding in the elliptic case to the equation

$$-\Delta u = \gamma(x, y, u. \nabla u) - f(u) \tag{1.2}$$

and in the parabolic case to

$$u_t - \Delta u = \gamma(x. t. u. \nabla u) - f(u) \tag{1.3}$$

where γ is a known function of its independent variables, and the problem is to determine the function $f(u)$. Physically one can view this as attempting to find the functional form of an unknown temperature - dependent source term in either a steady state or time dependent heat flow problem. Later we shall look at the problem of finding other unknown functions in (1.1) that correspond to conductivities or specific heats. In fact these problems turn out to be very similar to the one for (1.2) or (1.3), using our approach. This indicates that the method is a reasonably general one, in the sense that it is applicable to a wide class of equations although one may have to make some quite restrictive assumptions on the form of the data and a priori restrictions on the functions or coefficients to be determined. Finally we shall consider the case when (1.2) and (1.3) represent systems of elliptic and parabolic equations.

Our approach can be outlined as follows. One evaluates each term of the differential equation on that portion of the boundary where the overspecified data is prescribed and, using this data, obtains a mapping from the space of undetermined coefficients to the overposed data. In many cases solving the inverse problem can be shown to be equivalent to finding a fixed point of the mapping. In addition. with suitable constraints on the allowable solution class and the data, the mapping is contractive, and hence, in theory, the approach should lead to a constructive method of solution by an iterative procedure.

In the paper [8], the authors have given sufficient conditions under which one can obtain existence and uniqueness results for the problem of determining the function $f(u)$ from the parabolic equation (1.1), assuming that $\gamma(x. t. u, \nabla u)$ is known. A similar analysis was carried out for the elliptic equation (1.2) in [9].

It is the purpose of this talk to survey these results and in particular describe their numerical implementation. Our plan is to first summarize the mathematical theory underlying the iteration scheme, and then present numerical examples that describe the main features of the algorithm. For this part we shall concentrate on the problem of finding the unknown forcing function $f(u)$, given the rest of the differential equation.

The following discussion will be primarily descriptive. We shall avoid giving precise technical statements regarding assumptions on the the size and regularity of the data, instead referring to the papers 8-10, for the details. In this same spirit, we shall restrict our discussion to equations that are in the simplest form possible, consistent with retaining the main features of the problem at hand, and only introduce the other known terms and coefficients when these play a significant role in the speed of convergence of the numerical scheme.

The main idea.

In order to simplify the exposition, we shall for the present restrict our attention to the one spatial variable parabolic equation

$$u_t - u_{xx} = f(u) + \gamma(x,t,u,u_x) \tag{2.1}$$

in the region $0 < x < 1$, $0 < t < T$ for some $T > 0$, and the elliptic equation

$$-\Delta u = f(u) + \gamma(x,y,u,\nabla u) \tag{2.2}$$

defined in a bounded subset of the plane. Here the function $f(u)$ is unknown, except for the a priori assumption that it is a function of u alone. The remainder of the reaction term, γ, is a known function of its arguments. We may thus view $f(u)$ as a nonlinear perturbation of the general forcing term.

We require that the solution of (2.1) satisfy the primary boundary conditions

$$u_x(0,t) = g_0(t) \qquad u_x(1,t) = g_1(t) \qquad 0 \le t \le T \tag{2.3}$$

$$u(x,0) = 0 \qquad 0 \le x \le 1 \tag{2.4}$$

The prescribed overposed data, shall be the value of the temperature u on the boundary $x = 0$.

$$u(0,t) = \theta(t) \tag{2.5}$$

For the elliptic case we assume that (2.2) is defined in a bounded region Ω of $\mathbf{R}^2$, with smooth boundary $\partial\Omega$, and that $\partial\Omega$ can be decomposed into two connected components, $\partial\Omega = \partial\Omega_1 \cup \partial\Omega_2$. On $\partial\Omega_1$ we impose the boundary condition

$$\alpha\frac{\partial u}{\partial\nu} + \beta u = g_1 \qquad (x,y) \in \partial\Omega_1 \qquad (2.6)$$

while on $\partial\Omega_2$ we give *both* Dirichlet *and* Neumann boundary data

$$\begin{aligned} \frac{\partial u}{\partial\nu} &= g_2 \\ u &= \theta \end{aligned} \qquad (x,y) \in \partial\Omega_2 \qquad (2.7)$$

The domain is illustrated in fig .1. Here ν is the normal and σ is the tangent vector to the boundary curve $\partial\Omega$.

In both problems we seek to recover the solution pair $< u, f >$ from the combination of primary and overposed data.

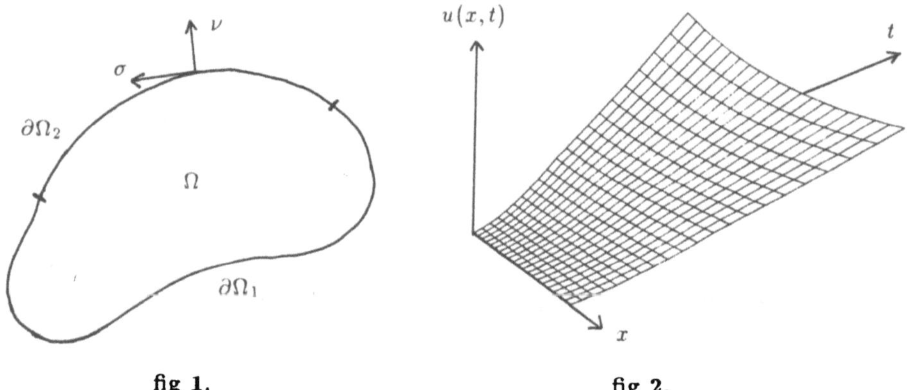

fig 1. fig 2.

One remark should be made immediately. It is necessary to have the solution u of the direct problem take on its maximum range on that portion of the boundary where the overposed data is prescribed in order to be able to recover $f(u)$ over the entire range of values of u that appear in equations (2.1), (2.2). This can be guaranteed by restricting the choice of data and using the maximum principle. For the parabolic equation, for example, we can force this condition by taking $a_0(t) > g_1(t) > 0$ and $\gamma_x(x,t,u) \leq 0$.

This corresponds to the physical situation of putting in more heat at the boundary $x = 0$ than at the boundary $x = 1$ and where the heat sources are greater the nearer one gets to the left hand boundary. The solution of the forwards problem will thus have the characteristic profile shown in fig.2. The desired condition can be achieved in the elliptic case. if α and β in (2.6) are non negative. the function g_2 in (2.7) is non negative and $\gamma = 0$.

If we evaluate the equation (2.1) on the boundary $x = 0$ then we obtain the following identity

$$\begin{aligned} f\big(\theta(t)\big) &= \theta'(t) - \gamma\big(0,t,\theta(t).g_0(t)\big) - u_{xx}(0,t;f) \\ &= \theta'(t) - \gamma\big(0,t,\theta(t).g_0(t)\big) - D_\nu u \end{aligned} \tag{2.8}$$

where we write $u(x,t;f)$ to denote the value at the point (x,t) of the solution of the forwards problem for a prescribed function f, and $D_\nu u$ to denote $u_{xx}(0,t;f)$.

Let us denote by $\mathbf{T}_\theta[\ ]$ the operator on the right hand side of (2.8), that is

$$\mathbf{T}_\theta[f] = \theta'(t) - \gamma\big(0,t,\theta(t),g_0(t)\big) - D_\nu u \tag{2.9}$$

It can be shown that $< u, f(u) >$ is a solution pair to the inverse problem if and only if f is a solution of (2.8),i.e. a fixed point of $\mathbf{T}_\theta[f]$ in the sense that $\mathbf{T}_\theta[f](\cdot) = f\big(\theta(\cdot)\big)$, a condition we shall refer to as a *theta fixed point*. Recall that we assume that $u(0,t) = \theta(t)$ is monotonic in t so that one can recover the function $f(\cdot)$ from knowledge of $f\big(\theta(t)\big)$.

For the elliptic boundary value problem (2.2),(2.6) and (2.7), if σ and ν denote unit vectors in the tangential and normal directions on the boundary $\partial\Omega$, then for some functions $a(s)$ and $b(s)$ of the arc length s. and with $a^2 + b^2 = 1$,

$$\frac{\partial}{\partial x} = a\,\frac{\partial}{\partial \sigma} + b\,\frac{\partial}{\partial \nu} \qquad\qquad \frac{\partial}{\partial y} = b\,\frac{\partial}{\partial \sigma} - a\,\frac{\partial}{\partial \nu}$$

Thus in this coordinate system the Laplacian has principal part $\dfrac{\partial^2}{\partial \sigma^2} + \dfrac{\partial^2}{\partial \nu^2}$. Let D_ν and D_σ denote the normal and tangential components of the Laplacian on the boundary $\partial\Omega$, that is

$$D_\nu = \frac{\partial^2}{\partial \nu^2} + c(s)\,\frac{\partial}{\partial \nu}$$

and

$$D_\sigma = \frac{\partial^2}{\partial \sigma^2} + d(s)\,\frac{\partial}{\partial \sigma}$$

where $c(s)$ and $d(s)$ are coefficients that depend only on the boundary $\partial\Omega$. (For example if Ω is the unit circle then $D_\nu = \frac{\partial^2}{\partial r^2} + \frac{1}{r}\frac{\partial}{\partial r}$ and $D_\sigma = \frac{1}{r^2}\frac{\partial^2}{\partial\theta^2}$).

If we evaluate each term of equation (2.2) on $\partial\Omega_2$ then we are led to following equation for the unknown $f(u)$

$$\mathbf{T}_\theta[f] \equiv -\tilde{\gamma}(s) - D_\sigma\theta - D_\nu\{u(x,y;f)\} = \hat{f}(\theta) \tag{2.10}$$

where $\hat{v}$ denotes the restriction of a function v to the boundary $\partial\Omega_2$, and $u(x,y;f)$ denotes the corresponding solution of the direct problem.

In both the elliptic and parabolic case we can thus represent the mapping from the unknown $f(u)$ to the boundary data in the form

$$\mathbf{T}_\theta[f](\cdot) = f\big(\theta(\cdot)\big) \tag{2.11}$$

Given all of this, a constructive algorithm for determining the fixed point is therefore available - namely to make an initial guess $f^{(0)}$ and to obtain successive iterates $f^{(n)}$ by the scheme

$$f^{(n+1)}\big(\theta(\cdot)\big) = \mathbf{T}_\theta[f^{(n)}](\cdot) \tag{2.12}$$

Note that each iterate requires that one can solve the primary problem with the given source term $f^{(n)}(u)$ in order to obtain the value of $u(\cdot,\cdot;f^{(n)})$ and hence $\mathbf{T}_\theta[f^{(n)}]$.

The standard a priori estimates for partial differential equations of elliptic and parabolic type imply that

$$\|D_\nu\{u(x,y;f)\}\|_{k,\alpha} \le C\|f\|_{k,\alpha} \tag{2.13}$$

for some constant C, and hence the iteration (2.12) is always defined. C depends on the domains, as well as primary and overposed data.

Note that it is essential to the above argument that the unknown reaction term should be a function of a single variable. This however, need not be u, but could be ∇u or indeed any a priori known function of u and ∇u, since both these quantities are given at the point $x_0 \in \partial\Omega_2$ as primary and overposed data.

Numerical implementation.

Our procedure can be outlined as follows

(a) A known function $f = f_{act}(u)$ was selected, and the differential equation was solved with $f(u) = f_{act}(u)$ and the primary boundary conditions to obtain the solution of the direct problem $u(\cdot, \cdot)$ numerically. The values of the function $\theta(\cdot)$, that is the values of u on the overposed boundary at a discrete set of points on this boundary, was then read off from the direct solution. This forms the analog of "experimental" data.

(b) These values of the function θ were then passed to the inversion routine as the overposed data. This routine implemented the iteration scheme outlined in the previous section: an initial guess $f = f_0$ was chosen and the direct problem was run to obtain the values of D_ν on the overposed boundary, and this was used in the update algorithm (2.12) to form the mapping $\mathbf{T}_\theta[\ ]$ and to obtain the next iterate.

In all computations of the direct problem or forwards problem for the parabolic equation we used a Crank-Nicholson scheme to perform the time integrations. For the elliptic equation we used an SOR iteration scheme.

In practice we ran the forwards solver to obtain the actual overposed data on a fairly course grid, in part to imitate experimental uncertainty. The discrete values of $\theta(\cdot)$ were then interpolated by a cubic spline (with optional smoothing). The resulting data formed the input to the inverse solver. Each of the successive approximations in the inversion routine requires the solving of a direct problem with an updated value of the unknown forcing function given by (2.12), and in order to minimize the error in this phase of the algorithm we chose a fine grid for the finite difference schemes.

The iteration scheme was terminated either after a fixed number of iterations or by a stopping condition, usually after the value of the supremum norm $\|u\|_{\partial\Omega} - \theta(\cdot)\|_\infty$ failed to decrease. Here $u|_{\partial\Omega}$ is the value of this function evaluated on the overposed boundary. Since $\theta(\cdot)$ is the only overposed data, in practice only this last measurement would be available for an error criterion. We shall call this the *standard error criterion*.

The values of the successive forcing functions $f_n(u)$ were stored in an array, and all interpolations between arrays of data were performed using a smoothing spline.

The above ideas are summarized in the flowchart. fig 3.

fig 3.   Algorithm Flowchart

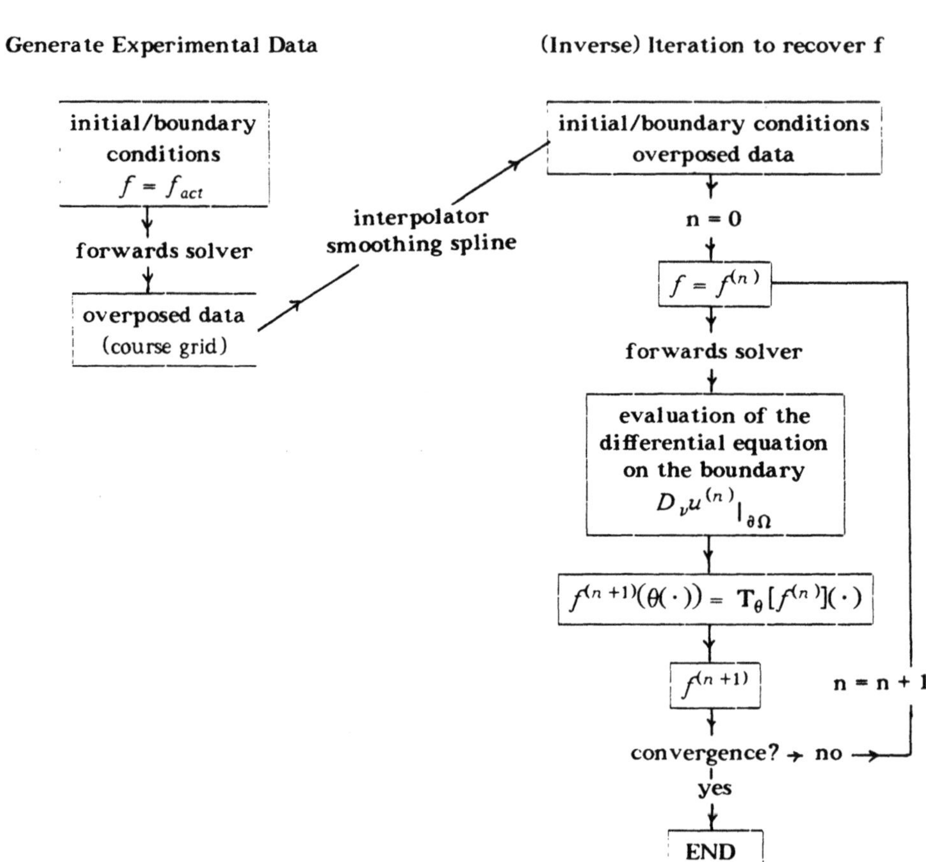

Let us demonstrate the iteration scheme by showing the results of an actual numerical calculation for the parabolic equation (2.1). We shall choose as a target function $f_{act} = 1 - \sin(2u)$, and take $f_0(u) = 0$ as the initial guess. This function is neither monotone nor convex, and increasing forcing functions are recovered more rapidly and more accurately by the algorithm than non monotone functions. In addition, numerical evidence supports the conjecture that the algorithm gives successive approximations that converge monotonically to the target function in this case. We choose the primary boundary conditions and the value of γ in order to obtain a range of temperatures that encompassed the main features of the target function. Taking $\gamma = 1$ and the values of the flux at $x = 0$ and $x = 1$ to be $u(0,t) = g_0(t) = 2t$ and $u(1,t) = g_1(t) = 0$.

Despite the condition of small time imposed by the results in [8], we had no problems recovering the forcing function $f(u)$ for arbitrary large times. The results are shown in fig 4. and table 1. In the table we give both the L_2 and the supremum norms of the difference of the nth iterate f_n and f_{act}, and the supremum and H_1 norms of the difference of the value of the nth iterated solution $u^{(n)}$ evaluated on the boundary $x = 0$ and the overposed data. One would normally be satisfied with convergence in the supremum norm, but the results of [8] guarantee convergence in a higher norm. In the graph we show the first and fifth iterates, f_1 and f_5 as well as f_{act}. In fig.4, the solid line represents the target function, and the two dashed lines represent the first and the fifth iterations.

Some additional smoothness on f and γ is required beyond continuity. Indeed, the global time existence of the direct problem is only guaranteed in general if the right hand side of (2.1) is uniformly Lipschitz in the variable u. Furthermore, the algorithm (2.12) requires the differentiability of the function $\theta(t)$. For this reason we include a norm of $u_n(0,t) - \theta(t)$ that contains information on higher derivatives of the boundary data. However the iteration scheme is able to recover, albeit with error at the points of irregularity, target functions that are much more poorly behaved, see [10] for an example of a discontinuous target function. The numerical schemes are in fact quite robust. Most of the theoretical limitations on a priori sizes of the norm of $f(u)$ do not seem to be numerical limitations in practice.

In general the number of iterations required for a given accuracy increases with an increase in the Lipschitz norm of the target function. An increase in the magnitude

table 1.

relative norms, $\|f_n - f_{act}\|/\|f_{act}\|$ **and actual norms,** $\|u_n(0,t) - \theta\|$
for the parabolic equation $u_t - u_{zz} = 1 + f(u)$

n	$\dfrac{\|f_n - f\|_\infty}{\|f\|_\infty}$	$\dfrac{\|f_n - f\|_{L^2}}{\|f\|_{L^2}}$	$\|u_n(0,t) - \theta(t)\|_\infty$	$\|u_n(0,t) - \theta(t)\|_{H^1}$
0	1.000000	1.000000	0.658585	0.878555
1	0.349445	0.315657	0.137365	0.267211
2	0.316741	0.196232	0.068856	0.128718
3	0.226623	0.155471	0.027563	0.079133
4	0.132801	0.100635	0.010813	0.051128
5	0.070848	0.054145	0.008111	0.029097
6	0.035747	0.026173	0.004777	0.014807
7	0.022895	0.013692	0.002143	0.007412
8	0.020669	0.009215	0.000818	0.004495
9	0.014734	0.006817	0.000495	0.003472

fig 4.

graphs of f_{act}, f_1 **and** f_5 **for the equation** $u_t - u_{zz} = 1 + f(u)$

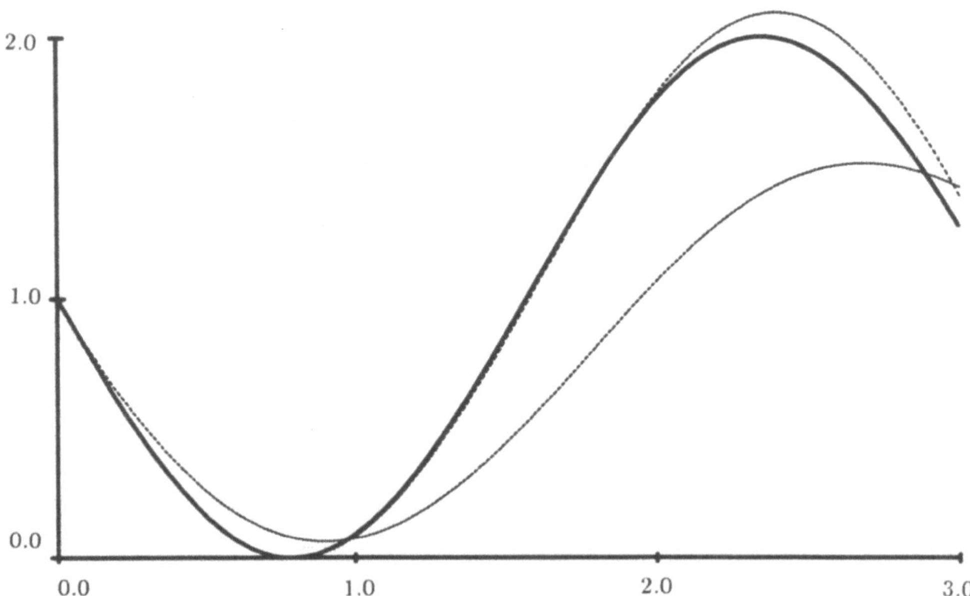

of the driving term γ and a decrease in the difference of g_0, g_1 increases the rate of convergence.

For the elliptic inverse problem we shall run the scheme for the equation $u_{xx} + u_{yy} = -f(u)$ defined on the square $\{ (x,y) : 0 < x < 1, \quad 0 < y < 1 \}$, taking as primary data

$$u(x,0) = x, \qquad u(1,y) = 1 + y, \qquad u(x,1) = 3 - x, \qquad u_x(0,y) = 0$$

and on $x = 0$ giving the overposed data $u(0,y) = \theta(y)$. The line $x = 0$ thus represents the boundary arc $\partial\Omega_2$ and the other three sides of the square represent $\partial\Omega_1$, which we can think of as starting at the point $(0,0)$ and going in a counterclockwise direction to the point $(0,1)$. Notice that on $\partial\Omega_1$ the Dirichlet data is a monotone function and that the maximum and minimum of $u(x,y)$ on $\partial\Omega$ occurs at the points $(0,1)$ and $(0,0)$. Once again we take $f_{act} = 1 - \sin(2u)$ as the target function, and $f_0(u) = 0$ as the initial guess. The results are shown in fig 5. and table 2.

The inversion routine is composed of the forwards solver plus interpolation, smoothing. and updating algorithms. There are several possible sources of error that can effect the scheme. The first comes from computational errors in the forwards solver routine, and a second comes from the smoothing routines used in handling the various arrays appearing in $\mathbf{T}_\theta$. These errors can be reduced by refining the grid sizes. The last two columns of tables 1 and 2 are essentially a measure of these errors. In fig.5 the solid line represents the target function, and the dashed lines represent the first and third iterations (the third iteration is indistinguishable on this scale from the target function).

More serious are the experimental errors generated in measuring $\theta(\cdot)$. The results in [8.9] require that this function posses derivatives that lie in some Hölder class, indeed if u is to be a strong solution to the problem certain regularity of the boundary values are inevitable. The first two columns of tables 1 and 2 measure the veracity (or reliability) of the data θ. If $\theta(\cdot)$ is noisy then one can of course use a smoothing spline but two problems in controlling this error should be mentioned.

(a) The update algorithm for the parabolic equation requires that one evaluate θ' and the update algorithm for the elliptic equation requires that θ'' be calculated. Most of the residual error ones sees in the above numerical runs stems from the calculation of these quantities, especially in the elliptic case. The problem is of course most noticeable at the endpoints of the interval.

table 2.

relative norms, $\|f_n - f_{act}\| / \|f_{act}\|$ and actual norms, $\|u_n(0,y) - \theta\|$
for the elliptic equation $-\Delta u = f(u)$

n	$\dfrac{\|f_n - f\|_\infty}{\|f\|_\infty}$	$\dfrac{\|f_n - f\|_{L^2}}{\|f\|_{L^2}}$	$\|u_n(0,y) - \theta(y)\|_\infty$	$\|u_n(0,y) - \theta(y)\|_{H^1}$
0	1.000000	1.000000	0.128807	0.310415
1	0.075779	0.081912	0.014061	0.032729
2	0.012502	0.010032	0.000644	0.001948
3	0.003458	0.003025	0.000226	0.000546
4	0.003739	0.001945	0.000116	0.000306
5	0.003674	0.001927	0.000116	0.000333

fig 5.

graphs of f_{act}, f_1 and f_3 for the equation $-\Delta u = f(u)$

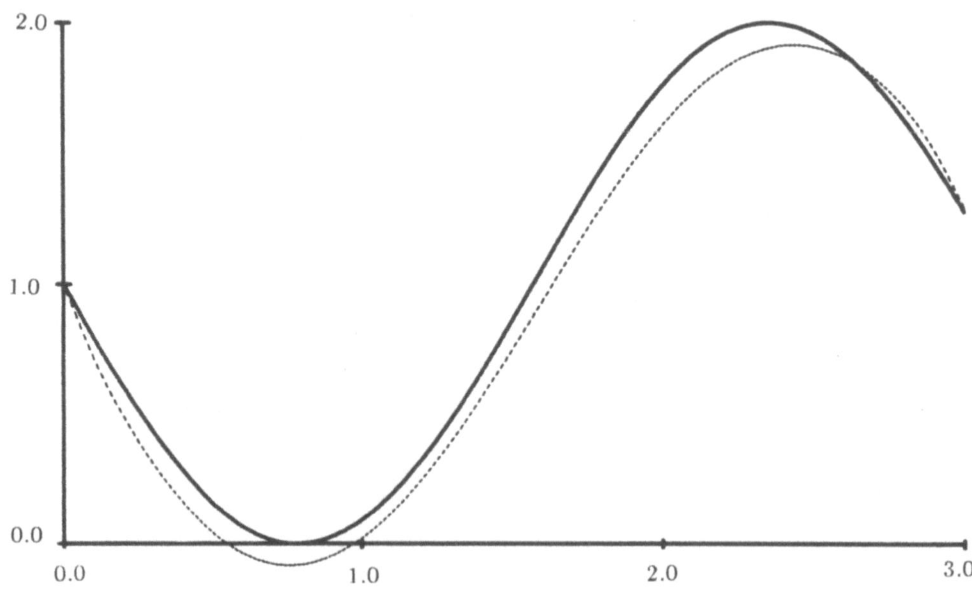

(b) A fairly large variation in the unknown forcing function gives rise in general to a much smaller differences in the values of u on the overposed boundary. This is evident from the tabulated data, in particular from table 2, where a fairly large variation in $f_{act} - f_0$ gives rise to smaller difference in $u_{act} - u_0$ that is, $\theta - u_0$ on the overposed boundary.

Other unknown coefficient problems.

Consider now the case where the source term is a known function of its independent variables, but either a conductivity or a specific heat has to be determined.

For the elliptic case we can consider the problem of finding both $u(x,y)$ and $a(u)$ in

$$-\nabla.a(u)\nabla u = 0$$

subject to mixed primary data

$$
\begin{aligned}
u &= g_1 & (x,y) &\in \partial\Omega_1 \\
a(u)\frac{\partial u}{\partial \nu} &= g_2 & (x,y) &\in \partial\Omega_2
\end{aligned}
\tag{4.1}
$$

and to the overposed condition

$$u = \theta \qquad\qquad (x,y) \in \partial\Omega_2 \tag{4.2}$$

By direct differentiation of the differential equation we can put (4.1) in the form

$$-\Delta u = b(u).|\nabla u|^2$$

where $b(u) = a'(u)/a(u)$. The problem can now be reduced to finding the theta fixed point of the mapping $\mathbf{T}_\theta[\cdot]$ defined by

$$\mathbf{T}_\theta b = \left\{ \frac{-D_\sigma\theta \cdot D_\nu\{u(x,y;b)\}}{(|\nabla u|^2)} \right\} \tag{4.3}$$

where the denominator $|\nabla u|^2$ is evaluated on $\partial\Omega_2$. This is well defined since the additional overposed boundary data yields the tangential and normal derivatives of u on the overposed boundary.

Of course certain obvious restrictions of uniformly positivity must be placed on $a(u)$ and the primary boundary data g_2, in addition to the same constraints of the maximum and minimum values of $u(x,y)$ occur on the arc $\partial\Omega_1$ Numerical implementation of the iteration scheme formed by this $\mathbf{T}_\theta[\;$ indicates that convergence is obtained in a similar manner to that of equation (2.2). Conditions under which this problem has at most one solution, as well as conditions under which there is non-uniqueness can be found in [11].

We can use a similar technique in the case of the parabolic equation

$$u_t - \left(a(u)u_x\right)_x = \gamma \tag{4.4}$$

By means of the change of variables $v = S_a(u) = \int_0^u a(r)\,dr$ we can transform this equation into

$$c(v)v_t - v_{xx} = \gamma \tag{4.5}$$

where $c(v) = a(u) = a\left(S_a^{-1}(v)\right)$

Equation (4.5) has interest in its own right, the function $c(u)$ corresponding to a temperature dependent specific heat. If we consider this equation with (2.3) as primary data and (2.4) as the overposed condition then we are led to an iteration scheme for the mapping $\mathbf{T}_\theta[c]$ defined by

$$\mathbf{T}_\theta[c] = \frac{\gamma(0,t,\theta(t),g_0(t)) + v_{xx}(0,t;c)}{\theta'(t)} \tag{4.6}$$

The conditions imposed on the primary data are similar to the case of unknown f; we must still ensure that the boundary on which the overposed data is given contains the entire range of values of $u(x,t)$, over the entire domain $\{\,(x,t):0 \le x \le 1,\ 0 \le t \le T\,\}$. We must also ensure that $u_t(0,0) \ne 0$, for in this case the value of $\theta'(0)$ would be zero and the mapping (4.6) would not be defined at the origin. This constraint can be met by requiring, for example, that $\gamma(0,0,0,0) \ne 0$. Details of the convergence of the iteration scheme defined by (4.5) can be found in [10].

Finally, we may extend this work to a system of coupled equations where the source terms are unknown. For example for parabolic equations in one space variable, we may wish to find the pair $<\vec{U},\vec{F}>$ in

$$\vec{U}_t - \vec{U}_{xx} = \vec{F}(\phi(\vec{U})) + \vec{\gamma}(x,t,\vec{U})$$

where

$$\vec{U} = \left(u^{(1)}, u^{(2)}, \ldots u^{(m)}\right)$$

$$\vec{F} = \left(f^{(1)}, f^{(2)}, \ldots f^{(m)}\right)$$

$$\vec{\gamma} = \left(\gamma^{(1)}, \gamma^{(2)}, \ldots \gamma^{(m)}\right)$$

and $\phi(\cdot)$ is a known function of $\vec{U}$. The function ϕ represents a known interaction term between the various components of $\vec{U}$. Possible choices of ϕ are

$$\phi(\vec{U}) = u^{(1)} u^{(2)} \ldots u^{(m)}$$

$$\phi(\vec{U}) = \sqrt{\left(u^{(1)}\right)^2 + \left(u^{(2)}\right)^2 + \ldots + \left(u^{(m)}\right)^2}$$

The primary boundary data would be of the form

$$u_x^{(i)}(0,t) = g_0^{(i)}(t) \qquad - u_x^{(i)}(1,t) = g_1^{(i)}(t) \qquad \text{for} \quad i = 1, 2, \ldots m$$

and the overposed data would be

$$u^{(i)}(0,t) = \theta^{(i)}(t) \qquad \text{for} \quad i = 1, 2, \ldots m$$

The mapping $\mathbf{T}_\theta[\ ]$ becomes in this case

$$\mathbf{T}_\theta[f_{n+1}^{(i)}] \equiv f_{n+1}^{(i)}(\Phi(t)) = \left(\theta^{(i)}(t)\right)' - \gamma^{(i)}\left(0, t, \theta^{(i)}(t)\right) - u_{xx}^{(i)}(0, t; f_n^{(i)}) \qquad (4.7)$$

with

$$\Phi(t) = \phi\left(\theta^{(1)}(t), \theta^{(2)}(t), \ldots \theta^{(m)}(t)\right)$$

The conditions that need to be imposed on the data are analogous to the case with one equation.

References.

1. J. R. Cannon *Determination of the unknown coefficient $k(u)$ in the equation $\nabla . k(u)\nabla = 0$ from overspecified boundary data*, J. Math. Anal. Appl. **18**, (1967), 112-114.

[2] J. R. Cannon and Paul C. DuChateau, *An inverse problem for a nonlinear diffusion equation*, S.I.A.M. J. Appl. Math., **39** (2), (1980), 272-289.

[3] Paul C. DuChateau , *Monotonicity and uniqueness results in identifying an unknown coefficient in a nonlinear diffusion equation* , S.I.A.M. J. Appl. Math., **41** (2), (1981), 310-323.

[4] Paul C. DuChateau and W. Rundell, *Unicity in an inverse problem for an unknown reaction term in a reaction diffusion equation*, J. Diff. Eqn. **59** (2), (1985), 155-164.

[5] G. Eriksson and G. Dahlquist, *On an inverse nonlinear diffusion problem*, Numerical Treatment of Inverse Problems in Differential and Integral Equations, Birkhäuser, Boston, 1983.

6. A. Lorenzi, *An inverse problem for a semilinear parabolic equation*, Annal. di Mat., Pura et Applica **82** (4), (1982). 145-166.

7. N. V. Muzylev, *Uniqueness theorems for some converse problems of heat conduction*, U. S. S. R. Comput. Maths. Math. Phys. **20**, (1980), 120-134.

8. M. Pilant and W. Rundell, *An Inverse Problem for a Nonlinear Parabolic Equation*. Comm. in P.D.E. **11** (4), (1986), 445-457.

[9] M. Pilant and W. Rundell, *An inverse problem for a nonlinear elliptic equation*, submitted.

[10] M. Pilant and W. Rundell, *Iteration schemes for unknown coefficient problems arising in parabolic equations*, submitted.

[11] M. Pilant and W. Rundell, *A uniqueness theorem for determining conductivity from overspecified boundary data*, submitted.

International Series of
Numerical Mathematics, Vol. 77
© 1986 Birkhäuser Verlag Basel

THE METHOD OF 'GENERALIZED INTERPOLATION'
FOR APPROXIMATE SOLUTION OF ILL-POSED PROBLEMS

Thomas I. Seidman

Abstract. Suppose we seek the solution $x_* \varepsilon X$ of an ill-posed problem which can be formulated as an infinite system of scalar equations: $\lambda_j(x) = \beta_j$ $(j = 1,2,\ldots)$. The method of generalized interpolation defines approximants x_N minimizing $\|x\|$ among solutions of the finite system: $\lambda_j(x) = \beta_j$ $(j = 1,\ldots,N)$. Under quite mild conditions one has $x_N \to x_*$, even when the method is modified for computational convenience.

1. INTRODUCTION

Suppose, for some set of points $\{t_1,\ldots,t_N\}$, one were given the values $\{x_*(t_1),\ldots,x_*(t_N)\}$. The standard problem of interpolation is to produce a function x_N such that $x_N(t_j) = \beta_j$ $(j = 1,\ldots,N)$ where $\{\beta_j\}$ is the set of given values. Of the various possible interpolating functions, the one selected might be characterized variationally. For example, the piecewise linear interpolant minimizes the integral of $|x'|^2$ while the (national) piecewise cubic spline interpolant achieves minimization of the integral of $|x''|^2$. One expects that as one increases the number of interpolating points one is improving accuracy so $x_N \to x_*$ in some suitable sense. Standard examples show that this need not be the case for, say, polynomial interpolation, but is true for the interpolation methods based on a variational selection principle[1].

We take an abstract view of this and consider the evaluations simply as functionals $\lambda_j : x \mapsto x(t_j)$ for which the sequence of values $\beta_j := \lambda_j(x_*)$ are available, i.e., x_* is uniquely determined (as $\{t_1,\ldots\}$ becomes dense) as the solution of the infinite system

---

[1] For example, if $x_* \varepsilon H^4(0,1)$ it is known that natural cubic spline interpolation (corresponding to $x_N'' = 0$ at 0, 1) provides uniform $\mathcal{O}(h^4)$ convergence except near the ends [2]. In this case we note that a modification of the variational selection, involving a more explicit concern for endpoint behavior, can give such uniform convergence on the entire interval [3] as the interpolating mesh becomes dense.

(1.1) $\qquad\qquad \lambda_j(x) = \beta_j \qquad (j = 1, \ldots)$

and the approximants x_N are determined by

(1.2) $\quad \|x_N\|_X = \min \quad$ subject to $\quad \lambda_j(x) = \beta_j \quad (j = 1, \ldots, N)$

We refer to (1.2) and variants as the <u>method</u> <u>of</u> <u>generalized</u> <u>inter-</u><u>polation</u> for (approximate) solution of problems formulated as in (1.1). In this generalized form we take X to be some specified Banach space in which the solution is known to lie and for which the functionals $\{\lambda_j\}$ are suitably continuous, no longer with any assumption that these take the form of point evaluations.

The discussion here is largely based on [7], [8].

2. BASIC CONVERGENCE THEOREM

In this section we wish to formulate specific hypotheses determining a useful class of situations for which one has assured convergence: $x_N \to x_*$ in X for a convenient variation of (1.2). We assume:

(H-1) X is a reflexive Banach space with the Efimov-Stečkin property: $x_n \rightharpoonup x$ with $\|x_n\| \to \|x\|$ implies $x_n \to x$ (e.g., X uniformly convex).

(H-2) $\{\lambda_j : j = 1, 2, \ldots\}$ is a sequence of functionals defined on a subset $B \subset X$; there is a unique $x_* \in S_* := \{x \in B: (1.1)\}$ minimizing $\|x - \tilde{x}\|$.

(H-3) For each N one has approximating functionals $\{\lambda_{j,N}: j = 1, \ldots, N\}$, approximate data $\{\beta_{j,N}: j = 1, \ldots, N\}$, and an approximating subset B_N; assume that $\beta_{j,N} \to \beta_j$ and that, for any subsequence $\{N = N(k)\}$ and corresponding $\{\hat{x}_N\}$ such that $\hat{x}_N \in B_N$ and $\hat{x}_N \to \hat{x}$, one has $\hat{x} \in B$ and
$\qquad$ (*) if $\lambda_{j,N}(\hat{x}_N) \to \beta_j$ then $\hat{\beta}_j = \lambda_j(\hat{x})$,

<u>Remark</u>: (H-3*) would follow if one knew that $\lambda_{j,N} \to \lambda_j$ uniformly on bounded sets and that each λ_j was continuous from the (sequential) weak topology.

$\quad$ THEOREM 2.1. <u>Assume</u> (H-1,2,3). <u>Suppose</u> $\varepsilon_{j,N} \to 0 \quad (N \to \infty)$ <u>with</u> $\varepsilon_{j,N}$ <u>large</u> <u>enough</u> <u>to</u> <u>ensure</u> (<u>for</u> <u>large</u> N) <u>that</u> $|\lambda_{j,N}(x_*) - \beta_{j,N}| \leq \varepsilon_{j,N}$ <u>so</u> x_* <u>is</u> <u>in</u> $S_N := \{x \in B_N: |\lambda_{j,N}(x) - \beta_{j,N}| \leq \varepsilon_{j,N}\};$ <u>set</u>

157

$$\nu_N := \underline{\inf}\,\{\|x - \tilde{x}\| : x \in S_N\} \quad \underline{\text{and}}\ \underline{\text{suppose}}\ \ 0 < \delta_N \to 0. \quad \underline{\text{Let}}\ \ x_N$$
$\underline{\text{be}}\ \underline{\text{any}}\ \underline{\text{solution}}\ \underline{\text{of}}$

(2.2) $x_N \in B_N: \quad |\lambda_{j,N}(x_N) - \beta_{j,N}| \leq \varepsilon_{j,N} \qquad (\underline{i}.\underline{e}.,\ \ x_N \in S_N);$

(2.3) $\|x_N - \tilde{x}\| \leq \nu_N + \delta_N.$

$\underline{\text{Then}}\ \ x_N \to x_* \ \underline{\text{as}}\ \ N \to \infty.$

$\underline{\text{Remark.}}$ (1.2) is just (2.2) with $B_N = X$, $\tilde{x} = 0$, $\lambda_{j,N} = \lambda_j$,
$\beta_{j,N} = \beta_j$, $\varepsilon_{j,N} = 0$, $\delta_N = 0$. The generalization here permits (i)
the use of imprecise observations $\beta_{j,N}$ and computational approx-
imations $\lambda_{j,N}$ to λ_j, (ii) approximate solution of the minimization problem (1.2), (iii) the use of $\tilde{x}$, B_N to involve any $\underline{a}$
$\underline{\text{priori}}$ information available about the solution.

$\underline{\text{Proof.}}$ Since $x_* \in S_N$ we have $\nu_N \leq \bar{\nu} := \|x_* - \tilde{x}\|$ whence $\{x_N\}$
is bounded. Extracting subsequences as necessary, we have
$x_N \to \bar{x}$. Since $\varepsilon_{j,N} \to 0$ and $\beta_{j,N} \to \beta_j$, we have $\lambda_{j,N}(x_N) \to \beta_J$
so (H-3*) gives $\lambda_j(\bar{x}) = \beta_j$ (for $j = 1,2,\ldots$); also, we have
$\bar{x} \in B$ so $\bar{x} \in S_*$. We have $\|\bar{x} - \tilde{x}\| \leq \lim \inf \|x_N - \tilde{x}\| \leq$
$\lim \sup \|x_N - \tilde{x}\| \leq \lim \sup(\nu_N + \delta_N) \leq \bar{\nu} = \|x_* - \tilde{x}\|$. By the unique-
ness of x_* as minimizer, this gives $\bar{x} = x_*$ and $\|x_N - \tilde{x}\| \to$
$\|x_* - \tilde{x}\|$ so (H-1) gives $x_N \to x_*$. Since this holds for each such
subsequence, it holds for the full sequence $\{x_N\}$ as desired. $\square$

3. DISCUSSION

We begin with the observation that (1.1) is an entirely natural way to formulate an inverse problem: each λ_j is a "measuring device" yielding the observed value β_j. The reasonable
assumption is that the infinite system (1.1) with "infinite pre-
cision" characterizes the desired solution x_* while, at any
"finite" time, one has available only a finite set of values
$\{\beta_{j,N} = \lambda_j(x_*) + [\text{small error}]: j = 1,\ldots,N\}$ with error estimates. One assumes that, with enough effort, one can take N as
large and make the measurements as precise as desired; similarly,
for computation one can approximate λ_j quite accurately by $\lambda_{j,N}$
and can then approximately solve the resulting approximation
(2.2), (2.3) to (1.2). The conclusion, then, is that use of this
approximation scheme ensures that one can get an arbitrarily good

approximation to x_* if one is willing to invest enough effort —
exactly as in the case of well-posed problems except that the
work required increases much more rapidly as the desired accuracy
is increased.

A recent example of application of this approach is to the
inverse Sturm-Liouville problem: given the eigenvalues $\lambda_j(\phi)$
for the operator

$$(3.1) \qquad \underset{\sim}{L} = \underset{\sim}{L}(\phi): y \mapsto -y'' + \phi y,$$

$$\mathcal{D}(\underset{\sim}{L}) = \{y: \underset{\sim}{L}y \in L^2(-1,1); y' = 0 \text{ at } -1,1\},$$

find ϕ. The sequence of eigenvalues (taken in increasing numerical order) gives (1.1); assuming ϕ is even on $[-1,1]$ one has
uniqueness so we might take, e.g., $B = \{\phi \geq 0: \text{even}\}$. Thus, the
approximant ϕ_N minimizes some norm while matching the first N
eigenvalues (for a computational approximation to $\underset{\sim}{L}(\phi)$ used for
the eigenvalue calculation). To admit, e.g., positive measures
(mass distributions) for ϕ, this was done using the H^{-1}-norm.
The most difficult part of the analysis, then, was showing that
(3.1) makes sense for $\phi \in H^{-1}$ and that weak-H^{-1} convergence for
a sequence $\{\phi_k\}$ gives convergence of the j-th eigenvalue to
obtain (H-3). See [9].

The original development of this approach [1], [4], [5] considered linear problems

$$(3.2) \qquad \underset{\sim}{A}x = b$$

in a Hilbert space setting $\underset{\sim}{A}: X \to Y$ with an <u>a priori</u> assumption
that $b \in \mathcal{R}(\underset{\sim}{A})$. Introducing a sequence $\{\eta_1, \eta_2, \ldots\}$ in Y^* $(= Y)$,
one has equivalence of (3.2) to

$$(3.3) \qquad \langle \lambda_j, x \rangle_X = \beta_j \qquad (j = 1,2,\ldots),$$

$$(3.4) \qquad \lambda_j := \underset{\sim}{A}^* \eta_j \in X^* \quad (= X),$$

$$(3.5) \qquad \beta_j := \langle \eta_j, b \rangle_Y$$

if $\{\eta_j\}$ is total on $\mathcal{R}(\underset{\sim}{A})$. Here the variational characterization (1.2) just gives $x_N \in X_N := \text{span}\{\lambda_1, \ldots, \lambda_N\}$ so $x_N = \sum c_k \lambda_N$

with the coefficients $\{c_k = c_{k,N}\}$ obtainable by solving[2] the $N \times N$ system:

$$(3.6) \qquad \sum_k H_{j,k} c_k = \beta_j \qquad (j = 1, \ldots, N)$$

where $(H_{j,k})$ is the Hessian matrix: $H_{j,k} := \langle \lambda_j, \lambda_k \rangle$. One then has convergence $x_N \to x_*$ as X_N "becomes dense" in the ortho-complement of $N(A)$. Given an <u>a priori</u> bound on $\|x_*\|_Z$ for some Z embedding compactly in X, one obtains [5] a convergence rate but otherwise the convergence can be arbitrarily slow — corresponding to convergence of the expansion of x_* with respect to the basis $\{e_j\}$ obtained by Gram-Schmidt orthonormalization of $\{\lambda_j\}$. For comparison, if one were to start with $\{e_j\}$ and the subspaces X_N and use the standard <u>least squares</u> approach

$$(3.7) \qquad \hat{x}_N \; \varepsilon \; X_N, \qquad \|A x_N - b\|_Y = \min,$$

then, even with exact knowledge of b and exact computation, examples show [6] that one need not have $x_N \to x_*$ but may actually have $\{x_N\}$ unbounded (or weakly, but not strongly, convergent to x_*). Note that (3.7) corresponds to the use of the "norm" on X induced by A^*A for minimization in (1.2) and that is equivalent to the X-norm only when the problem is well-posed. The minimization in (1.2) serves to stabilize (regularize) the computation — and, as modified in (2.3), to take advantage of a preliminary estimate.

It is interesting to distinguish between <u>a priori</u> and <u>a posteriori</u> approaches corresponding to a distinction between the questions:

>>   What (feasible) combination of experiment and analysis will produce an acceptable approximation to the desired solution?

>>   What is the best (most useful) interpretation which can be obtained from the available data?

In this context, Theorem 2.1 represents an <u>a priori</u> concern but the specific approximation scheme, obtaining x_N from (2.2),

---

[2] Since x_N will just be the orthogonal projection (in X) of x_* to the subspace X_N, the system always has a solution (assuming exact computation) even if $\{\lambda_1, \ldots, \lambda_N\}$ might be linearly dependent. We do expect (3.6) to be ill-conditioned for large N so practical questions intervene at this point for the implementation.

(2.3), is intended to address, to some extent, the <u>a posteriori</u> question. The choices of X (and its norm), $\tilde{x}$, B_N, and $\varepsilon_{j,N}$ represent the use of auxiliary prior information about the solution x_*, the relevant topology for the application, and the accuracy[3] of the available measurements. It is possible to generalize the approach further (e.g., using lsc convex functionals other than a norm for the minimization — see [7], [8] — and so permit a more flexible treatment of various forms of prior information.

---

[3] We have considered here only the context of error bounds rather than a statistical treatment of the errors for which our treatment suggests a modified Bayesian approach.

REFERENCES

1. W. Chewning and T.I. Seidman, A convergent scheme for boundary controls for the heat equation. SIAM J. Control. Opt. 15 (1977) 64-72.

2. D. Kershaw, A note on the convergence of interpolatory cubic splines. SIAM J. Numer. Anal. 8 (1971) 67-74.

3. R. Korsan and T.I. Seidman, Endpoint formulas for interpolatory cubic splines. Math. Comp. 26 (1972) 897-900.

4. T.I. Seidman, The solution of singular equations, I: linear equations in Hilbert space, Pac. J. Math. 61 (1975) 513-520.

5. T.I. Seidman, Computational approaches to ill-posed problems: general considerations. Proc. 1976 Conf. on Inf. Sci., Systems, pp. 258-262, Johns Hopkins Univ. (1976).

6. T.I. Seidman, Nonconvergence results for the application of least squares estimation to ill-posed problems, J. Opt. Th. Appl. 30 (1980) 535-547.

7. T.I. Seidman, Convergent approximation methods for ill-posed problems, Part I: general theory. Control and Cybernetics 10 (1981) 31-49.

8. T.I. Seidman, Convergent approximation methods for ill-posed problems, Part II: applications. Control and Cybernetics 10 (1981) 51-71.

9. T.I. Seidman, A convergent approximation scheme for the inverse Sturm-Liouville problem. Inverse Problems 1 (1985) 251-262.

Thomas I. Seidman
Department of Mathematics
University of Maryland Baltimore County
Catonsville, Maryland 21228
U.S.A.   BITNET: seidman@umbc

Acknowledgments. The preparation of this paper was partially supported under grant # AFOSR-82-0271 based on research done with the support of that grant and also ARO support under DAAG-29-77-G-0061.
Thanks are due to the organizers of this conference both for the opportunity to include this paper and for having organized a most stimulating conference.

International Series of
Numerical Mathematics, Vol . 77
© 1986 Birkhäuser Verlag Basel

Gel'fan-Levitan's theory and related inverse problems

Takashi Suzuki

Introduction. Gel'fand-Levitan's theory [1] is the inverse theory of that of Titchmarsh-Kodaira [2]. The latter constructs the spectral density function $\rho = \rho(\lambda)$ from the differential operator $-\dfrac{d^2}{dx^2} + p(x)$ on $(0,\infty)$ with $(-\dfrac{d}{dx} + h)\cdot|_{x=0} = 0$, while the former reconstructs (p,h) from ρ. Thus, both theories are not only constructive, but also apply to singular boundary value problems. In the present article, we describe some remarkable structures of Gel'fand-Levitan's theory and their applications to the study of identifiability[1] for partial differential equations. This kind of works were reported in [3], [4] and [5], but some advances have been done after that. New ideas will be picked up.

§1. Gel'fand-Levitan's theory. In what follows, we are interested in the regular case. For $P = (p,h,H) \in X \equiv C^1[0,1] \times \mathbf{R} \times \mathbf{R}$, A_P denotes the Sturm-Liouville operator $-\dfrac{d^2}{dx^2} + p(x)$ with the boundary condition $(-\dfrac{d}{dx} + h)|_{x=0} = (\dfrac{d}{dx} + H)\cdot|_{x=1} = 0$. Then, its spectrum $\sigma(A_P)$ consists of simple eigenvalues:

$\sigma(A_P) = \{\lambda_n(P)\}_{n=0}^{\infty}$ $(-\infty < \lambda_0(P) < \lambda_1(P) <\cdots \to +\infty)$. Henceforth, $\varphi_n = \varphi_n(\cdot;P)$ denotes the eigenfunction of A_P corresponding to $\lambda_n = \lambda_n(P)$ and normalized by $\varphi_n|_{x=0} = 1$. Putting $\rho_n(P)$

---

1) Identifiability is the notion of uniqueness and stability in
    inverse problems.

$$= \int_0^1 \varphi_n(x;P)^2 dx,$$ we call $\{\rho_n(P)\}_{x=0}^{\infty}$ the norming constants. Also, the set of two sequences $\mathcal{C}(P) = \{\lambda_n(P), \rho_n(P)\}_{n=0}^{\infty}$ is called the spectral characteristics of A_p. For this case, Gel'fand-Levitan's theory asserts that P is recovered from $\mathcal{C}(P)$.

In particular, for $P, Q \in X$ the relation $\mathcal{C}(P) = \mathcal{C}(Q)$ implies $P \equiv Q$. This fact is taken up by M. M. Lavrentiev and K. G. Reznickaja ([6]) to show uniqueness in a hyperbolic inverse problem. Another application was given by A. Pierce ([7]). He considered the parabolic equation

$$(1) \qquad \frac{\partial u}{\partial t} + \left(- \frac{\partial^2}{\partial x^2} + p(x)\right) u = 0 \qquad (0 < x < 1, \quad 0 < t < T)$$

under the boundary condition

$$\left(- \frac{\partial}{\partial x} + h\right)u\big|_{x=0} = 0, \qquad \left(\frac{\partial}{\partial x} + H\right)u\big|_{x=1} = F \qquad (0 < t < T)$$

and the initial condition

$$u\big|_{t=0} = 0 \qquad (0 < x < 1).$$

He showed that in the case of $F \neq 0$, the functions $F = F(t)$ and $f = f(t) \equiv u\big|_{x=1}$ $(0 \le t \le T)$ determine the spectral characterisitcs $\mathcal{C}(P)$ and hence $P = (p,h,H) \in X$.

§2. <u>Integral Tansformation</u>. The reconstruction due to [1] is described as follows. For $p \in C^1[0,1]$, $h \in R$ and $\lambda \in \mathbb{R}$, let $\varphi = \varphi(x;p,h,\lambda)$ be the solutions of

$$\left(- \frac{d^2}{dx^2} + p(x)\right)\varphi = \lambda\varphi \quad (0 \le x \le 1) \quad \text{with} \quad \varphi\big|_{x=0} = 1, \quad \varphi'\big|_{x=0} = h.$$

By the definition, we have $\varphi_n(\cdot;P) = \varphi(\cdot;p,h,\lambda_n(P))$. Then, the relation

$$(2') \qquad \varphi(x;p,h,\lambda) = \cos\sqrt{\lambda}x + \int_0^t K(x,y)\cos\sqrt{\lambda}y\,dy \qquad (0 \le x \le 1)$$

is derived, where $K = K(x,y) \in C^2(\bar{D})$ $(D = \{(x,y)\,|\,0 < y < x < 1\})$ is independent of $\lambda \in \mathbb{R}$ and solves the hyperbolic equation

$$K_{xx} - K_{yy} = p(x)K \qquad ((x,y) \in D)$$

with

$$K(x,x) = h + \frac{1}{2}\int_0^x p(s)\,ds, \qquad K_y(x,0) = 0 \qquad (0 \le x \le 1).$$

From Parseval's relation, the equality

$$(3) \qquad F(x,y) + \int_0^x K(x,z)F(z,y)\,dz + K(x,y) = 0 \qquad ((x,y) \in D)$$

follows, which we call the Gel'fand-Levitan equation. Here, $F = F(x,y)$ is determined only by $\mathcal{C}(P)$ and the unique solvability of (3) with respect to $K = K(t,y)$ holds under appropriate assumptions for $\mathcal{C}(P)$. Therefore, in order to reconstruct $P = (p,h,H)$ form $\mathcal{C} = \mathcal{C}(P)$, we have only to solve (3). Then, $P = (p,h,H)$ is recovered as

$$p = 2\frac{d}{dx}K(x,x), \quad h = K(0,0) \quad \text{and} \quad H = -\mathcal{G}'(1;p,h,\lambda_n(P))/\mathcal{G}(1;p,h,\lambda_n(P)).$$

For more details, see [5].

Many authors have modified these relations ([8], [9], [10]). For instance, it is almost straightforward that (2') extends like

$$(2) \qquad \mathcal{G}(x;q,j;\lambda) = \mathcal{G}(x;p,h,\lambda) + \int_0^x K(x,y)\mathcal{G}(y;p,h,\lambda)\,dy \quad (0 \le x \le 1).$$

where $K(x,y) = K(x,y;q,j;p,h)$ is defined through

$$(4.1) \qquad K_{xx} - K_{yy} + p(y)K = q(x)K \qquad ((x,y) \in D)$$

with

$$(4.2) \qquad K(x,x) = (j-h) + \frac{1}{2}\int_0^x (q(s)-P(s))\,ds, \qquad K_y(x,0) = hK(x,0)$$
$$(0 \le x \le 1).$$

These relations are made use of in the study of identifiability for the parabolic equation (1) under

$$(-\frac{\partial}{\partial x} + h)u\big|_{x=0} = (\frac{\partial}{\partial x} + H)u\big|_{x=1} = 0 \quad \text{and} \quad u\big|_{t=0} = a(x).$$

Here, $P = (p,h,H) \in X$ and $a \in L^2(0,1)$ are unknown and to be

determined by, for instance, $f_j = u|_{x=j}$ $(j = 0,1; 0 \leq t \leq T)$.
See [4] and also Murayama [11].

To study the problem to determine P, a from $g_1 = u|_{x=x_0}$, $g_2 = \frac{\partial}{\partial x} u|_{x=x_0}$ $(0 \leq t \leq T)$, for instance, another modification as (2') is necessary. For this problem, a sensitive dependence on the point x_0 exists for identifiability to hold. Such a phenomenon is also observed for similar inverse problems for parabolic equations on the circle. These results are reported in [4], and we do not go into details.

More direct relations hold for the solution $u = u(x,t)$ of the parabolic equation. For instance, if $u = u(x,t)$ solves (1) with

$$(- \frac{\partial}{\partial x} + h)u|_{x=0} = 0,$$

then, for $K = K(x,y)$ defined through (4), the function

$$v(x,t) = u(x,t) + \int_0^x K(x,y)u(y,t)dy$$

solves

$$\frac{\partial v}{\partial t} + (- \frac{\partial^2}{\partial x^2} + q(x))v = 0 \qquad (0 < x < 1, \ 0 < t < T)$$

with

$$(- \frac{\partial}{\partial x} + j)v|_{x=0} = 0 \quad \text{and} \quad v|_{x=0} = u|_{x=0} \qquad (0 < t < T).$$

W. Rundell made use of this fact to get a constructive standpoint in a parabolic inverse problem ([12]). It is no wonder that a similar fact holds for hyperbolic equations, which has been made use of in the study of "stability" of a hyperbolic inverse problem (Suzuki [13]). (This "stability" does not mean the well-posedness of the problem). See also [14].

§3. __Duality and Symmetry.__ For $p \in C^1[0,1]$ and $\lambda \in \mathbb{R}$, let $\varphi = \varphi(x;p,\infty,\lambda)$ be the solution of

$$(- \frac{d^2}{dx^2} + p(x))\varphi = \lambda\varphi \ (0 \leq x \leq 1) \quad \text{with} \quad \varphi|_{x=0} = 0, \quad \varphi'|_{x=0} = 1.$$

167

Then, the relation

(5) $\varphi(x;q,\infty,\lambda) = \varphi(x;p,\infty,\lambda) + \int_0^x K(x,y)\varphi(y;p,\infty,\lambda)dy \quad (0 \le x \le 1)$

holds, wher $K = K(x,y;q,\infty;p,\infty) \in C^2(\bar{D})$ is independent of
λ and solves the hyperbolic equation (4.1) with

(4.2') $K(x,x) = \frac{1}{2}\int_0^x (q(S) - p(S))ds, \quad K(x,0) = 0 \quad (0 \le x \le 1).$

In the formula (2), we considered different (but of the same
kind) boundary condition at $x = 0$, namely, $\varphi|_{x=0}=1, \; \varphi'|_{x=0}=h$
and $\psi|_{x=0} = 1, \; \psi'|_{x=0} = j$. Here in this formula, we are con-
cerned with the same boundary conditions $\varphi|_{x=0} = 0, \; \varphi'|_{x=0} = 1$
and $\psi|_{x=0} = 0, \; \psi'|_{x=0} = 1,$ but the relation (5) similar to (2)
holds. This is because the second relation gives $K(0,0) = 0$
as a compatibility condition in (4.2'). This slight difference
yields, for instance, the following "duality" between A_p and
A_p^* (Suzuki [15]).

  Here, A_p^* denotes the Sturm-Liouville operator $-\dfrac{d^2}{dx^2}+ p(x)$
with $\cdot|_{x=0} = \cdot|_{x=1} = 0.$ Its eigenvalues are denoted by $\sigma(A_p^*)$
$= \{\lambda_n^*(p)\}_{n=1}^{\infty} \quad (-\infty < \lambda_p^*(p) < \lambda_2^*(p) < \cdots \to + \infty).$ Put

$C_S^1 = \{q \in C^1[0.1] | q(1 - x) = q(x) \quad (0 \le x \le 1)\}.$ Then,

<u>Proposition</u>. For $p,q \in C_S^1$ and $n_1 \in N^* \equiv \{1,2,\ldots\},$ there
exists an $h \in \mathbb{R}$ such that

$$\lambda_n((p,h,h)) = \lambda_n((q,h,h)) \quad (n \ne n_1)$$

if and only if

$$\lambda_n^*(p) = \lambda_n^*(q) \quad (n \ne n_1).$$

Furthermore, in case of $q \ne p,$ we have

$$\lambda_{n_1}((p,h,h)) = \lambda_{n_1}^*(q) \quad \text{and} \quad \lambda_{n_1}^*(p) = \lambda_{n_1}((q,h,h)). \quad \square$$

This proposition could be a key in the study of inverse Sturm-Liouville problem ([15]).

Another kind of symmetry is also seen in Gel'fand-Levitan's theory. Let $\varphi^* = \varphi^*(x;P,H,\lambda)$ be the solution of

$$(-\frac{d^2}{dx^2} + p(x))\varphi^* = \lambda\varphi^* \quad (0 \le x \le 1) \quad \text{with} \quad \varphi^*|_{x=1} = 1, \ \varphi^{*\,'}|_{x=1} = H.$$

Then, in the same way as in (2), the relation

$$\varphi^*(x;q,J,\lambda) = \varphi^*(x;p,H,\lambda) + \int_x^1 K^*(x,y)\varphi^*(y;p,H,\lambda)dy$$

holds true, with an appropriate kernel $K^* = K^*(x,y;q,J;p,H)$ $\in C^2(\overline{D^*})$, where $D^* = \{(x,y)\,|\,0 < x < y < 1\}$. In way of this fact, K. Iwasaki gave a different derivation of the Gel'fand-Levitan equation (3) from making use of Parseval's relation ([16]). Namely, let $\mathcal{K}$ and $\mathcal{K}^*$ be the integral operators on $[0,x]$ and $[x,1]$ with the kernels $K(x,\cdot)$ and $K^*(x,\cdot)$, respectively:

$$(\mathcal{K}g)(x) = \int_0^x K(x,y)g(y)dy, \quad (\mathcal{K}^*g)(x) = \int_x^1 K^*(x,y)g(y)dy.$$

An integral operator $\mathcal{F}$ on $[0,1]$ is defined through

$$1d + \mathcal{F} = (1d + \mathcal{K})^{-1}(1d + \mathcal{K}^*).$$

Then, the kernel $F = F(x,y)$; $(\mathcal{F}g)(x) = \int_0^1 F(x,y)g(y)dy$ is C^2 in D and D^*, and satisfies (3) in D, for instance. Importantly, this way of derivation of (3) gives a direct calculation of the gap $F(x - 0,x) - F(x + 0,x)$ at the diagonal of the kernel, that is,

$$F(x-0,x) - F(x+0,x) = C_0(Q) - C_0(P) \quad (P=(p,h,H),\ Q=(q,j,J)),$$

where $C_0(P) = h + H + \frac{1}{2}\int_0^1 p(S)dS$ and $C_0(Q) = j + J + \frac{1}{2}\int_0^1 q(S)dS$. The value $C_0(P)$ appears, for instance in connection with the asymptotic behavior of $\lambda_n(P)$:

169

$$\lambda_n(P)^{\frac{1}{2}} = n\pi + C_0(P)/n + o(\tfrac{1}{n}) \qquad (\text{as } n \to \infty).$$

Let $\mathscr{F}_x$ be the integral operator on $[0,x] \times [0,x]$ associated with the kernel $F(x,\cdot)$. In the case of $C_0(Q) = C_0(P)$, $F(x,\cdot)$ becomes continuous on $[0,x] \times [0,x]$. Then, the solution $K = K(x,\cdot)$ of the Gel'fand-Levitan equation (3) can be expressed in terms of the Fredholm determinant $\tau(x) = \det(1d + \mathscr{F}_x)$. In particular, we have

$$K(x,x) = -\frac{d}{dx} \log \tau(x) \qquad \text{so that} \qquad q = p - 2\frac{d^2}{dx^2} \log \tau(x).$$

These formulas are also useful in the study of inverse Sturm-Liouville problem [(16)].

§4. <u>Iso-spectral Deformation</u>. Two operators are said to be iso-spectral if they have a same spectral set. Iso-spectral deformation has been studied by the method of "iso-spectral flow" by H. P, McKean, E. Trubowitz and others, which would be described in J. R. McLaughlin's article. Here, we wish to present another approach from the view point of Gel'fand-Levitan's theory. The motivation is to extend our study on the identifiability for partial differential equations to multi-dimensional cases.

To begin with, we see that (2) reads;

$$(6) \qquad \mathscr{G}(x;q,j,\lambda) = \mathscr{G}(x;p,h,\lambda) + \int_0^1 H(\lambda \cdot y)K(x,y)\mathscr{G}(y;p,h,\lambda)\,dy.$$

Here $H = H(z)$ is the Heaviside function: $H(z) = \begin{cases} 1 & (z > 0) \\ 0 & (z < 0) \end{cases}$.

Singularity of Heaviside function H will produce δ-functions when its commutator with the Laplacian, $[\frac{d^2}{dx^2}, H]$, is taken.

Since there is no natural way of extending H to multi-dimensional spaces, it seems to be impossible to extend (2) for those cases. However, under the assumption of iso-spectral: $\sigma(A_P) = \sigma(A_Q)$, (6) is nothing but

$$(7) \qquad \mathscr{G}_n(\cdot,Q) = (1d + \mathscr{K})\mathscr{G}_n(\cdot;P),$$

170

where $\mathcal{K}$ is the integral operator with the kernel $H(x-y)K(x,y)$.
We note that conversely, the operator $\mathcal{K}$ can be defined through
(7):

$$\mathcal{K}: \varphi_n(\cdot;P) \longmapsto \varphi_n(\cdot;Q) - \varphi_n(\cdot;P) \qquad (n = 0,1,2,\ldots).$$

The corresponding kernel $\tilde{K} = \tilde{K}(x,y)$ is formally given as

$$(8) \qquad \tilde{K}(x,y) = \sum_{n=0}^{\infty} \{\varphi_n(x;Q) - \varphi_n(x;P)\}\varphi_n(y;P)/\rho_n(P).$$

From this standpoint, (2) is nothing but

$$(9) \qquad \tilde{K}(x,y) = H(x - y)K(x,y).$$

The formal expression (8) will have meanings even in multi-dimensions, through (9) does not. That idea is made use of in
[17]. There, from a study of the ultra-hyperbolic operator
$-\Delta_x + \Delta_y + p(x) - q(y)$, we have shown the uniqueness in a parabolic inverse problem of multi-space dimension.

REFERENCES

1.   I. M. Gel'fand and B. M. Levitan, AMS Transl. (2) 1 (1955),
     253-304.

2.   E. C. Titchmarsh, Eigenfunction Expansion Associated with
     Second-order Differential Equations, Oxford Univ. Press,
     London, 1939.

3.   T. Suzuki, In; Computing Methods in Applied Science and
     Engineering, V, (R. Glowinski, J. L. Lions, eds.), Noth-
     Holland Amsterdam, 1982.

4.   T. Suzuki, In; Nonlinear Partial Differential Equations in
     Applied Science, (H. Fujita, P. D. Lax, G. Strang, eds.),
     North-Holland, Amsterdam, 1983.

5.   T. Suzuki, J. Fac. Sci. Univ. Tokyo, Sect. IA 32 (1985),
     223-271.

6.   M. M. Lavrentiev, K. G. Reznickaja, Soviet Math. Dokl., 20
     (1979), 1416-1418.

7.   A. Pierce, SIAM J. Control Optim, 17 (1979), 494-499.

8.   D. L. Colton, Solution of Boundary Value Problems by the Method of Integral Transformation, Pitman, London/San Fransisco/Melbourne, 1976.

9.   R. Carroll, Transmutation Theory, North-Holland, Amsterdam, 1985.

10. C. Kravaris, J. H. Seinfeld, to appear.

11. R. Murayama, F. Fac. Sci. Univ. Tokyo, Sect. IA, 28 (1981), 317-330.

12. W. Rundell, Rocky Mountain J., Math., 13 (1983), 679-688.

13. T. Suzuki, Proc. Japan Acad. Ser. A. 56 (1980), 259-263.

14. T. Suzuki, In; Proc. of the IMACS International Symposium on Medelling and Simulation for Control of Lumped and Distributed Parameter Systems in Lille, 1986.

15. T. Suzuki, J. Differential Equations, 56 (1985), 165-194.

16. K. Iwasaki, to appear.

17. T. Suzuki, Proc. Japan Acad., 62. Ser. A (1986), 83-86.

Takashi Suzuki
Department of Mathematics
Faculty of Science
University of Tokyo

International Series of
Numerical Mathematics, Vol. 77
© 1986 Birkhäuser Verlag Basel

RECONSTRUCTION AMBIGUITIES OF INVERSE
SCATTERING ON THE LINE

P.C. SABATIER

Abstract. We study one dimensional scattering problems governed by the Schrödinger equation or by the impedance equation. In the absence of bound states, there exists a class of potentials (resp. impedances) that is bijectively related with a class of spectral data, ie reflection coefficients as a function of energy. On the other hand, it is known in larger classes several examples of different potentials that are consistent with a given reflection coefficient and no true bound state. In the lecture below it is shown that

(1) these ambiguities are related with a Darboux-type transformation which is defined on very wide classes of potentials (resp. impedances), leaves invariant the Schrödinger (resp. impedance) equation whereas the reflection coefficient is flipped, and depends on one arbitrary parameter, so that we obtain a one parameter family of "equivalent" potentials (resp. impedances).

(2) If potentials classes (resp. classes of impedance factors) are defined by their leading asymptotic behavior $[\ell_{+} (\ell_{+}+1)]x^{-2}$, resp. (x^{p},x^{q}), the transformation takes a potential (resp. an impedance factor) from one class to another one.

(3) The transformation leaves the transmission coefficient invariant but introduces or suppresses zero energy bound states or half bound states, so that it is *not* isospectral.

Generalizations are suggested in the end of the lecture.

Foreword. This lecture contains a survey of some results recently obtained and partially published by P.C. Sabatier [1,2] and A. Degasperis & P.C. Sabatier [3].

1. INTRODUCTION AND BACKGROUND

The one-dimensional Schrödinger equation

$$(1.1) \qquad f'' + (k^2 - u)f = 0 \qquad (f, f' \text{ continuous})$$

admits for potentials u belonging to a wide class, hereafter called the scattering class V, solutions that are defined for real k by their asymptotic behavior as x tends to infinity, the so-called Jost solutions :

$$(1.2) \qquad f_+(k,x) \sim \exp[i\,k\,x] \qquad x \to \infty$$

$$(1.3) \qquad f_-(k,x) \sim \exp[-i\,k\,x] \qquad x \to -\infty$$

which are related with each other by the formula

$$(1.4) \qquad T(k)\, f_-(k,x) = f_+(-k,x) + R^+(k)\, f_+(k,x)$$

This formula defines the transmission coefficient $T(k)$ and the reflection coefficient $R^+(k)$.

The impedance equation

$$(1.5) \qquad \left(\alpha^{-2} \frac{d}{dx}\, \alpha^2 \frac{d}{dx} + k^2\right) p(k,x) = 0 \qquad (p,\ \alpha^2 \frac{dp}{dx} \text{ continuous})$$

admits for impedance factors α belonging to a wide class, hereafter called A, solutions that are defined for real k by their asymptotic behavior :

$$(1.6) \qquad p_+(k,x) \sim \alpha^{-1} \exp[i\,k\,x] \qquad (x \to +\infty)$$

$$(1.7) \qquad p_-(k,x) \sim \alpha^{-1} \exp[-i\,k\,x] \qquad (x \to -\infty)$$

and are related by the formula

$$(1.8) \qquad T(k)\, p_-(k,x) = p_+(-k,x) + R^+(k)\, p_+(k,x)$$

which defines the transmission and reflection coefficients.

For both equations the direct problem is to construct $R^+(k)$ from u, resp. α, in V, resp. A. The inverse problem aims at going back from $R^+(k)$ to u, resp. α. These problems have been thoroughly studied [4,5] in the subclass L_1^1 of v :

$$(1.9) \qquad L_1^1 = \{u \mid \int_{-\infty}^{+\infty} (1+|x|)\,|u(x)|\,dx < \infty\}$$

In particular, it has been proved [6] that the inverse problem $R^+ \to u$ is well-posed if

    (a)   there is no value $k^2 = -k_n^2$, $k_n^2 \geq 0$, for which the equation (1.1) has a solution in $L^2(\mathbb{R})$ (so called "bound state")

    (b) u is sought in the subclass L_2^1 of L_1^1 :

$$(1.10) \qquad L_2^1 = \{u \mid \int_{-\infty}^{+\infty} (1+x^2)\, |u(x)|\, dx < \infty\}$$

    (c) $R^+(k)$ is in the image R of L_2^1. This image has been characterized [6]. We only notice here that for $R^+ \in R$, if there is no zero energy bound state, $R^+(k)$ must go to -1 as k goes to zero.

The impedance equation (1.5) is equivalent to the Schrödinger equation (1.1) if α is real, with absolutely continuous derivative (ie with a second derivative locally integrable). The equivalence is achieved by setting $f = \alpha p$, $u = \alpha''/\alpha$, whereas $T(k)$ and $R^+(k)$ are the same for both equations. Hence, there corresponds to L_2^1 (with no bound state) a class of impedance factors, called the regular ones, such that the inverse problem $R \to \alpha''/\alpha$ is well posed from R to this class. In some sense, this class is easy to identify because the positivity of α is sufficient to prevent a bound state.

The impedance equation still holds if α and $\dfrac{d\alpha}{dx}$ are only piecewise absolutely continuous. We can allow a finite number of "singular points" forming the set

$$(1.11) \qquad S = \{x_o,\ x_1,\ x_2, \ldots, x_N\}$$

where α and α' may have jumps (exclusive of other singularities). Let us define the "transmission factors", "reflection factors", "slope factors" at these points :

$$(1.12) \qquad t_n^{-1} = \frac{1}{2}\left(\frac{\alpha_n^+}{\alpha_n^-} + \frac{\alpha_n^-}{\alpha_n^+}\right)$$

$$(1.13) \qquad r_n = \frac{1}{2}\, t_n \left(\frac{\alpha_n^+}{\alpha_n^-} - \frac{\alpha_n^-}{\alpha_n^+}\right)$$

$$(1.14) \qquad s_n = \frac{1}{2}\, t_n \left(\frac{{\alpha'}_n^-}{\alpha_n^+} - \frac{{\alpha'}_n^+}{\alpha_n^-}\right)$$

where α_n^+ stands for $\alpha(x_n^+)$ etc. Notice that $t_n^2 + r_n^2 = 1$. The values of t_n, r_n, s_n, are called the "singular data". They are not sufficient to construct an impedance factor α from its asymptotic value and the so-called "background potential", ie the locally integrable function equal to α''/α at all non singular point. One would also need for it the additional "singular data" :

$$(1.15) \qquad q_n = \frac{1}{2} t_n \left(\frac{\alpha'_n^-}{\alpha_n^+} + \frac{\alpha'_n^+}{\alpha_n^-} \right)$$

On the other hand, the q_n's are not necessary in a study of the inverse problem for the impedance equation because of the following theorem, which shows a "standard equivalence" between impedances [3].

<u>Theorem.</u> Let α and β be two piecewise differentiable positive functions such that β/α is continuous and $W = \alpha\beta' - \beta\alpha'$. Then, if p is a solution of (1.5), $q = \alpha p/\beta$ is a solution of the equation obtained from (1.5) if α is replaced by β.

Thanks to this theorem [3] the "standard equivalent" impedances are described by two constants , W and C :

$$(1.16) \qquad \beta(x) = W\,\alpha(x) \int^x \frac{dt}{\alpha^2(t)} + C\,\alpha(x)$$

If we do not care of the arbitrary multiplicative constant, we see that the study of this equivalence reduces to setting first $W = 1$, C arbitrary, then $W = 0$, $C = 1$. In the transformation (1.16) from α to β, $T(k)$, $R(k)$, t_n, r_n, s_n, are invariant, q_n is not. In our study of the inverse problem, we shall always consider that two "standard equivalent" impedance factors need not be identified. Hence, if α is regular and positive, the inverse problem is well-posed from R to the corresponding class of impedances.

Thus we know a class of potentials and a class of impedances for which the inverse problem from $R^+(k)$ in R is well-posed. On the other hand, examples of non uniqueness in the one-dimensional Schrödinger inverse scattering were given in the last few years. Let us cite only two of them :

(1) For the following scattering data :

$$(1.17) \qquad R^+(k) = (1-ik)^{-1} \qquad T(k) = -ik(1-ik)^{-1}$$

there exists at least three equivalent potentials :

$$(1.18a) \qquad u_1(x) = -2\,\delta(x) + 8(2x+1)^{-2}\,\theta(x)$$

$$(1.18b) \qquad u_2(x) = u_1(-x)$$

$$(1.18c) \qquad u_3(x) = -2\,\delta(x) + 2(1+|x|)^{-2}$$

where $\theta(x)$ is the Heaviside function. The examples (a) and (b) were given by Abraham, de Facio, and Moses [7], and (c) was by Brownstein [8]. Brownstein [8] gave also the following example :

$$(1.19) \qquad R^+(k) = \tfrac{1}{2}\,i\,(k^2+3)\,\{(k+i)\,[(k+i)(k+\tfrac{1}{2}i) - 1]\}^{-1}$$

$$(1.20a) \qquad u_1(x) = -\delta(x) + 6(x+1)^{-2}\,\theta(x)$$

$$(1.20b) \qquad u_2(x) = -\delta(x) + 6(x-1)^{-2}\,\theta(-x)$$

In these examples, the potentials are not in L^1_2. First, of course, we notice the δ-function. It corresponds to one singular point in the impedance equation. Now the direct problem for the impedance equation with a finite number of singular point has been dealt with in a form that relates the scattering data of the impedance and those of the background potential [2,3]. A term $-2\,s_o\,\delta$ gives a slope singularity s_o at the point $x_o = 0$, with $r_o = 0$, $t_o = 1$. Hence there is a way of analysis of the δ-function effect, but we can hardly expect it yields an explanation of the non uniqueness, since a δ function is a local limit of L^1_2 functions. A much more serious departure from L^1_2 is the x^{-2} behavior at ∞. On the other hand, we notice the behavior of $R^+(k)$ which in the first examples goes to $+1$ as $x \to 0$. Had we written down the transmission coefficient in the second example, another remark would have been done : a $O(k^3)$ behavior at $k \to 0$.

In the second part of the lecture, we shall map the class of "regular" impedance factors, resp. potentials, into the largest class A, resp. V of admissible impedance factors, resp. potentials, by means of an algebraic transformation which is the limit of a Darboux transformation. Since convenient Darboux transformation would introduce non uniqueness by introducing bound states, it is no surprising to see that their limit introduces non uniqueness with or without introducing a zero-energy bound state.

In the third part of the lecture, we shall try some enlargements or generalizations of this theory.

2. CONSTRUCTION OF AMBIGUITIES

The reflection-flip transformation

Let α be a given impedance factor, β be defined from α by a convenient choice of W and C in (1.16), and let us set

$$(2.1) \qquad \tilde{\alpha} = \beta^{-1}$$

It is easily checked that

$$(2.2) \qquad p^T(k,x) = p(k,x) - \alpha\beta\, W^{-1}\, p'(k,x)$$

is a solution, with p^T and $\tilde{\alpha}^2\,\dfrac{dp^T}{dx}$ continuous, of

$$(2.3) \qquad [\tilde{\alpha}^{-2}\,\frac{d}{dx}\,\tilde{\alpha}^2\,\frac{d}{dx} + k^2]\, p^T(k,x) = 0$$

The singular points are unchanged. The asymptotic behavior of p_+^T and p_-^T is

$$(2.4) \qquad \beta\, p_+^T(k,x) \sim - ik\, W^{-1}\, \exp[i\,k\,x]$$

$$(2.5) \qquad \beta\, p_-^T(k,x) \sim ik\, W^{-1}\, \exp[-i\,k\,x]$$

so that the new Jost solutions are

$$(2.6) \qquad \tilde{p}_+(k,x) = (-i\,k)^{-1}\, p_+^T(k,x)$$

$$(2.7) \qquad \tilde{p}_-(k,x) = (i\,k)^{-1}\, p_-^T(k,x)$$

and the new scattering coefficients are related with the former ones by

$$(2.8) \qquad \tilde{T}(k) = T(k)$$

$$(2.9) \qquad \tilde{R}^+(k) = - R^+(k)$$

The singular data are also transformed according to the rule

$$(2.10) \qquad \tilde{t}_n = t_n$$

$$(2.11) \qquad \tilde{r}_n = - r_n$$

$$(2.12) \qquad \tilde{s}_n = - t_n^{-2}\, (1 + r_n^2\, s_n - 2\, t_n^{-2}\, r_n\, q_n)$$

$$(2.13) \qquad \tilde{q}_n = - t_n^{-2} (1+r_n^2)q_n - 2 t_n^{-2} r_n s_n$$

As we already noticed, the transformation can be studied without loss of generality, by considering first $W = 1$, so what

$$(2.14) \qquad \overset{\sim}{\alpha}(x) = [\alpha(x)]^{-1} [\int_0^x \alpha^{-2}(t) \, dt + c]^{-1}$$

and $\tilde{p}_+$, $\tilde{p}_-$ are given by (2.6), (2.7) and (2.2), considering in a separate step $W = 0$, $c = 1$, so that $\overset{\sim}{\alpha}(x)$ is simply $[\alpha(x)]^{-1}$ and the limit form of (2.2) is

$$(2.15) \qquad p^T(k,x) = \alpha^2 \, p'(k,x)$$

We shall call this separate case the "special transformation". From the impedance factor $\overset{\sim}{\alpha}$, and if there is no singular point (resp. if there are only "weak" singularities, ie $\tilde{r}_n = 0$, $\tilde{t}_n = 1$), we can construct a potential, which is a function of x equal to $\overset{\sim}{\alpha}''/\overset{\sim}{\alpha}$ (resp. this function between the singular points, and δ-measures of weight $- 2 \tilde{s}_n$ at the singular points). The potential is a regular measure at any finite point of $\mathbb{R}$, unless $\overset{\sim}{\alpha}$ happens to vanish, usually yielding then a double pole to the potential. Because of this double pole, the scattering problem is cut into two parts, and convenient continuations of solutions through the double pole must be settled. It is commonly accepted that the scattering problem has no meaning, because simple physical remarks are not able to fix the proper continuation. We personnally think that the physical relevance of these "ghosts" is still an open problem, but of course we shall provisionnally rank them among unacceptable potentials, and so we shall do for the corresponding impedances.

The ambiguities

The general transformation (2.14) depends on the arbitrary parameter c . Hence, all the derived impedance factors $\overset{\sim}{\alpha}(c,x)$, together with α^{-1} (that can be associated with $c = \infty$), correspond to the same reflection coefficient $- R^+(k)$. $T(k)$ being unchanged, no true bound state may have been created by the transformation. Thus, provided they belong to the scattering class A, the new $\overset{\sim}{\alpha}(C,x)$ form a one parameter class of ambiguities. So do the corresponding potentials $\overset{\sim}{\alpha}''/\overset{\sim}{\alpha}$, provided they belong to the "scattering class" V. The class L_2^1, resp. the corresponding class of impedances, are transformed, and the new classes contain potentials, resp. impedances, with asymptotic behaviors that are not allowed in L_2^1, resp. in the corresponding class of impedances. Let $P(\ell_-, \ell_+)$ be the class of potentials u such that

$$(2.16) \quad \left. \begin{array}{l} u(x) - \ell_-(\ell_-+1)\, x^{-2} \in L_1^1(-a, -\infty) \\[2ex] u(x) - \ell_+(\ell_++1)\, x^{-2} \in L_1^1(a, +\infty) \end{array} \right\} P(\ell_-,\ell_+) \quad (a > 0)$$

Without loss of generality, we can limit our study to non negative ℓ_-, ℓ_+. The class $L_1^1(\mathbb{R})$ corresponds to $\ell_- = \ell_+ = 0$. Now let $[p,q]$ be the class of impedance factors asymptotic to $C\, x^p$ as $x \to -\infty$, and to $C'\, x^q$ as $x \to +\infty$, where C, C' are unspecified positive constantes, p, q are non negative. For $u \in L_1^1(\mathbb{R})$, it is not difficult to prove that the impedance factor is generally a linear combination of one in $[1, 0]$ and one in $[0, 1]$. There are however special potentials such that α goes to a limit at $x \to \pm\infty$. It has been proved [3] that α in then the Jost solution corresponding to a zero energy bound state of this potential (Hint : notice that $\int_{-\infty}^{+\infty} (\alpha'^2 + V\,\alpha^2)\, dx = 0$). Indeed, there also exists in this case a zero energy bound state of the impedance equation, since $p = 1$ obviously satisfies

$$(2.17) \quad \left\{ \begin{array}{l} \alpha^{-2}\, \dfrac{d}{dx}\, \alpha^2\, \dfrac{dp}{dx} = 0 \\[3ex] \int_{-\infty}^{+\infty} dx\ \alpha^2\, \left(\dfrac{dp}{dx}\right)^2 = 0\ . \end{array} \right.$$

If a potential u is asymptotic to $p(p-1)\, x^{-2}$ as $x \to +\infty$, there exists an impedance factor α such that $\alpha''/\alpha = u$, and going to infinity like $C\, x^p$, and another impedance factor $\underline{\alpha}$ obtained from the first one by the formula $\alpha(x) \int_x^\infty \alpha^{-2}(t)\, dt$ and going to zero like $C'\, x^{-p-1}$. Same results hold at $-\infty$. "Transforming" the impedance factors α and $\underline{\alpha}$ yields impedance factor $\tilde{\alpha}$ and $\underline{\tilde{\alpha}}$ which have almost everywhere an absolutely continuous derivative. The asymptotic behavior of the corresponding potentials is readily derived from that of the impedance factors in this sense that $q(q-1)\, x^{-2}$ correspond to x^q. Hence, one can derive the table of transformations of potential classes from that of transformations of impedance classes and we shall give only the latter one.

$$(2.18) \quad [0,0] \begin{array}{l} \xrightarrow{\ *\ } [0,\, 0] \\[2ex] \searrow [-1,\, -1]\ \text{with a pole on}\ \mathbb{R} \end{array}$$

$$
(2.19) \quad [1,\ 0]
\begin{cases}
\overset{*}{\nearrow}\ [-1,\ 0] \\[4pt]
\rightarrow [-1,\ -1]\ \text{with a pole on } \mathbb{R},\ \text{if}\quad c < \int_{-\infty}^{0} dt/\alpha^2(t) \\[4pt]
\rightarrow [0,\ -1]\quad \text{if } c = \int_{-\infty}^{0} dt/\alpha^2(t) \\[4pt]
\rightarrow [-1,\ -1]\ \text{if } c > \int_{-\infty}^{0} dt/\alpha^2(t)
\end{cases}
$$

$$
(2.20) \quad [0,\ 1]
\begin{cases}
\overset{*}{\nearrow}\ [0,\ -1] \\[4pt]
\rightarrow [-1,\ -1]\quad \text{if}\quad c < -\int_{0}^{\infty} dt/\alpha^2(t) \\[4pt]
\rightarrow [-1,\ 0]\quad \text{if}\quad c = -\int_{0}^{\infty} dt/\alpha^2(t) \\[4pt]
\rightarrow [-1,\ -1]\ \text{with a pole on } \mathbb{R},\ \text{if}\quad c > -\int_{0}^{\infty} dt/\alpha^2(t)
\end{cases}
$$

if p, q are non negative integers

$$
(2.21) \quad [-p,\ -q]
\begin{cases}
\overset{*}{\nearrow}\ [p,\ q] \\[4pt]
\searrow [-(p+1),\ -(q+1)]\ \text{with a pole on } \mathbb{R}
\end{cases}
$$

If $p,\ q$, are integers ≥ 1

$$
(2.22) \quad [p,\ q]
\begin{cases}
\overset{*}{\nearrow}\ [-p,\ -q] \\[4pt]
\rightarrow [-p,\ -q]\ \text{if}\quad c < -\int_{0}^{\infty} dt/\alpha^2(t) \\[4pt]
\rightarrow [-p,\ q-1]\ \text{if}\quad c = -\int_{0}^{\infty} dt/\alpha^2(t) \\[4pt]
\rightarrow [-p,\ -q]\ \text{with a pole on } \mathbb{R},\ \text{if}\quad -\int_{0}^{\infty}\frac{dt}{\alpha^2(t)} < c < \int_{-\infty}^{0}\frac{dt}{\alpha^2(t)} \\[4pt]
\rightarrow [p-1,\ -q]\ \text{if}\quad c = \int_{-\infty}^{0} dt/\alpha^2(t) \\[4pt]
\rightarrow [-p,\ -q]\ \text{if}\quad c > \int_{-\infty}^{0} dt/\alpha^2(t)
\end{cases}
$$

One should notice the values which break down the classes of impedances (or potentials) into unphysical ones and "ghosts".

Zero energy bound states

Although most results could be generalized to classes $\bar{P}(\ell_-,\ \ell_+)$ with but a few complications, we now stick at subclasses $P(\ell_-,\ \ell_+)$ that contain the potentials u such that

$$
(2.23) \qquad u(x) = \frac{\ell_-(\ell_-+1)}{(x-a_-)^2}\,\theta(x_- - x) + \frac{\ell_+(\ell_++1)}{(x-a_+)^2}\,\theta(x - x_+) + \bar{u}(x)
$$

with $x_- < a_-,\ x_+ > a_+$, and $\bar{u}(x)$ such that

182

(2.24) $\quad\quad\quad \forall\ m \geq 0,\ \int_{-\infty}^{+\infty} dx\ |x|^m\ |\bar{u}(x)| < \infty$

Thanks to these assumption, it is possible to define the solutions of (1.1) which are analytic functions of k^2 and have the following limit dehavior :

$$(2.25)\quad\quad \sigma^{\pm}(x,k) = \frac{(x_- a_{\pm})^{\ell_{\pm}+1}}{(2\ell_{\pm}+1)\ !!} + O(k^2)$$

$$(2.26)\quad\quad \gamma^{\pm}(x,k) = (2\ell_{\pm}-1)\ !!\ (a_{\pm}-x)^{-\ell_{\pm}} + O(k^2)$$

These solutions are exact combinations of the Jost solutions, well defined for $k \neq 0$, and they remain themselves well defined as k goes to zero. Hence, it is possible to define from them a matrix Q, which is related to the usual problem S-matrix for any $k \neq 0$, and goes to a definite limit as k goes to zero. Q is given by the formula :

$$(2.27)\quad\quad \begin{bmatrix} \sigma^+(x,k) \\[2ex] \gamma^+(x,k) \end{bmatrix} = \begin{bmatrix} Q^{11}(k) & Q^{12}(k) \\[2ex] Q^{21}(k) & Q^{22}(k) \end{bmatrix} \begin{bmatrix} \sigma^-(x,k) \\[2ex] \gamma^-(x,k) \end{bmatrix}$$

One easily checks the following properties

$$(2.28)\quad\quad \begin{cases} \det Q = (-)^{\ell_+ + \ell_-} \\[1ex] Q^{ij}(-k) = Q^{ij}(k) \end{cases}$$

Setting $Q^{ij}(0) = q^{ij}$, we can use these limits to characterize the zero-energy behaviors of the transmission and reflection coefficients, together with the zero-energy spectrum, if any. The following theorems have been proved [3] :

<u>Theorem I</u> : $q^{21} = 0$ is the necessary and sufficient condition for the existence of a zero energy discrete eigenvalue. The corresponding solution of (1.1) is bounded and, if $\ell_+ \ell_- \neq 0$, it belongs to $L^2(\mathbb{R})$ ("true zero energy bound state").

<u>Theorem II</u> : The zero energy behavior of the scattering coefficients is given by the following formulas, where $R^-(k)$ is the reflection coefficient defined by the scattering of a wave arriving from the right, and equal to

$- R^+ (-k)$

(a) if there is no discrete eigenvalue at $k = 0$, ie $q^{21} \neq 0$:

(2.29)
$$\begin{cases} T(k) = (2/q^{21}) \, (i \, k)^{\ell_+ + \ell_- + 1} \, [1 + O(k)] \\ R^\pm(0) = (-)^{\ell_\pm + 1} \end{cases}$$

(b) if there is a discrete eigenvalue at $k = 0$ ie $q^{21} = 0$

(b 1) if both ℓ_+ and ℓ_- are larger or equal to 1.

(2.30)
$$\begin{cases} T(k) = (2/r) \, (i \, k)^{\ell_+ + \ell_- - 1} \, [1 + O(k)] \\ R^\pm(0) = (-)^{\ell_\pm + 1} \end{cases}$$

where r is a real number that does not need be identified

(b 2) if $\ell_+ = 0$, $\ell_- \geq 1$,

(2.31)
$$\begin{cases} T(k) = (2/q^{11}) \, (i \, k)^{\ell_-} \, [1 + O(k)] \\ R^+(0) = 1 \\ R^-(0) = (-)^{\ell_- + 1} \end{cases}$$

(b 3) if $\ell_+ \geq 1$, $\ell_- = 0$

(2.32)
$$\begin{cases} T(k) = (2/q^{22}) \, (i \, k)^{\ell_+} \, [1 + O(k)] \\ R^+(0) = (-)^{\ell_+ + 1} \\ R^-(0) = 1 \end{cases}$$

(b 4) $\ell_+ = \ell_- = 0$, $q^{22} = tg \, \alpha/2$

(2.33)
$$T(0) = \sin \alpha \quad R^\pm(0) = \pm \cos \alpha$$

Theorem III

Taking into account these remarks, it is not difficult to show that if a potential is such that $q^{21} \neq 0$, the limit transformation acting on it produces a class of equivalent potentials which usually have a zero energy discrete eigenvalue. If a potential is such that $q^{21} = 0$, the limit transformation usually either produces a ghost or suppresses the zero energy discrete eigenvalue.

For understanding this theorem, it must be kept in mind that the transformation maps a class $\bar{P}$ (defined by (2.16)) into another class $\bar{P}$ but <u>not</u> a class P (defined by (2.23),(2.24)) into another class P. Hence, even if we start from a class P, all the statements of theorem I and II are not generally valid after transformation, because the functions σ usually become undefined. However q^{21}, which is proportional to the wronskian of functions γ, is defined in any class $\bar{P}$ and keeps his properties.

Geometrical research of ambiguities

Let C be the set of parameters : $C = V$ if the parameters are potentials, $C = A$ if they are impedance factors. Let $E = R$ be the set of results (reflection coefficients). They are related with the parameters by means of (1.1), (1.4) or (1.5), (1.8), which define a mapping M from C to E. A way for solving the inverse problem is based on constructing joined routes in C and E (ie routes in E which are the M-images of routes in C). If the nets of these routes are sufficiently dense in C and E, they yield together two joined sets of generalized coordinates and any point in C or E can be approximated by a point with finitely many coordinates. If there is a continuous ambiguity, there should be a route in C corresponding to one point in C. Now, let T be a "geometrical transformation", ie a mapping of $C \times E$ into $C \times E$ which "commutes" with M, in this sense that if $x \in C$, $e = M(x) \in E$, and $T(x,e) = (\tilde{x}, \tilde{e})$, then $\tilde{e} = M(\tilde{x})$. It is clear that if T depends on a parameter a, two joined routes are defined by giving x_o, $x_e = M(x_o)$, and $T(a)$ say $x = X(x_o, a)$, $e = E(e_o, a)$.

We identified in V the set L_2^1, and its subset — say L^* — obtained by keeping only potentials without "bound state". Let R be the M-image of L^*. We know that R is also the M-image of L_2^1. The inverse problem is well-posed from R to L^*. From R to L_2^1, it is underdetermined : for each discrete eigenvalue there is an infinity of equivalent impedance factors depending on one parameter, the normalizing coefficient of the eigenvalues. Needless to say, these remarks, as well as those given below, can be transposed to the impedance formulation, but we would like avoiding to duplicate here analyses already written [3].

There exists in $\mathcal{V}$ a transformation of the kind described above. It is obtained by applying the so called "Darboux transformation" to the solutions of (1.1). Here we are only concerned with the result, which can be stated in this way. Let u be in $\mathcal{V}$, and let $f(k_o,x)$ be a solution of (1.1) for $k = k_o$. We can write down

$$(2.33) \qquad f(k_o,x) = \alpha f_-(k_o,x) + \beta f_-(-k_o, x)$$

Assume $\text{Im } k_o > 0$, $\alpha \neq \beta R_-(k_o)$, and k_o not a bound state. Then the transformation $T(k_o)$ is such that

$$(2.34) \qquad \begin{cases} \tilde{u}(x) = u(x) - 2 \dfrac{d}{dx} [f'(k_o,x)/f(k_o,x)] \\[2em] \tilde{R}^+(k) = \dfrac{k_o + k}{k_o - k} R^+(k) \end{cases}$$

Let $k_o = i\gamma$, $\gamma > 0$. The transformation depends on γ and on the arbitrary parameter α/β (one value being excluded), but, unless $\beta = 0$, the transformation modifies the transmission coefficient and introduces a bouns state at $k = i\gamma$. If u was in L^*, the parameter α/β is the hidden parameter describing the class of equivalent potentials which corresponds to one bound state at $k = i\gamma$. Hence, the underdetermination of the inverse problem from R to L_2^1 is enlightened by this geometrical transformation. By letting $\gamma \to 0^+$, we must be able either to see what happens on the boundary $L^* L_2^1$ (if the ambiguities disappear) or to get out of L_2^1 (if the ambiguities remain). Indeed, if $\gamma \to 0^+$, we obtain the limit reflection-flip transformation (this is obvious from 2.34), the transmission coefficient is unchanged by the transformation (this also could be readily obtained), but the transformation is not really isospectral, since it introduces a discrete zero energy eigenvalue, and it takes out of L_2^1.

3. GENERALIZATIONS

The geometrical research of ambiguities, or singularities, can be managed when transformations of the type we described are available and enable us to span C and E. There are several well-known cases in the litterature. Here I shall construct transformations which correspond to the very general problems described by Beals and Coifman systems, i.e. generalized Zacharov-Shabat systems [9]. Particular cases reduce these problems to those of Schrödinger equations or systems, with or without energy-dependent potentials [10], so that the transformations can also be reduced to these cases. Up to my knowledge, the general result given below is new. It shows a complete set of "elementary transformations" where the new parameters can be constructed. A general elementary transformation does not reduce to explicit formulas and is not a Darboux transformation. But if a scattering matrix is available, it is multiplied by a rational factor, if we deal with a spectral transform, it is added a Dirac measure — and thus an "elementary transformation" is a simple generalization of those described above.

The problem (ie the mapping M) is described by the equation

$$(3.1) \qquad \frac{d\Psi}{dx} = \xi\, J\, \Psi + Q\, \Psi$$

where ξ is a complex number, J, Q, and the unknown Ψ, are n×n matrices ; J is diagonal : $J_{ij} = \lambda_i\, \delta_{ij}$, $\lambda_i \neq \lambda_j \iff i \neq j$; Q is a zero-diagonal matrix $Q \in C = \{Q \mid \int_{-\infty}^{+\infty} \Sigma \mid Q_{ij}(x) \mid dx = \|Q\| < \infty\}$. E is the set of possible measurements either of a scattering matrix, if it can be defined, or of a spectral transform (see below). A transformation is defined by

$$(3.2) \qquad \tilde{\Psi} = T\ \Psi\ (T_\infty)^{-1}$$

where T depends on x, T_∞ does not the, tilda labels the transformed elements. It commutes with M iff $\tilde{\Psi}$ is a solution of (3.1) and hence iff

$$(3.3) \qquad \frac{\partial T}{\partial x} + (T\,Q - \tilde{Q}\,T) + \xi\,(T\,J - \tilde{J}\,T) = 0$$

We are only interested by the transformations which leave J invariant : $\tilde{J} = J$. "Trivial transformations" are those which do not depend on ξ . The condition (3.3) then shows that T is diagonal, constant, and

$$(3.4) \qquad \tilde{Q}_{ij} = Q_{ij}\, T_{ii}\ /\ T_{jj}$$

Let us now study ξ-linear transformations, $T = S + \xi U$. By cancelling the coefficients of ξ^2, ξ, ξ^o, in (3.3), we obtain

$$(3.5a) \qquad U \, J - J \, U = 0$$

$$(3.5b) \qquad \frac{\partial U}{\partial x} + U \, Q - \tilde{Q} \, U + S \, J - J \, S = 0$$

$$(3.5c) \qquad \frac{\partial S}{\partial x} + S \, Q - \tilde{Q} \, S = 0$$

Because all the λ_i's are different, U, which commutes with J, is diagonal. The diagonal terms in (3.5b) reduce to $\frac{\partial U}{\partial x}$ and hence U is constant.

In the following, I study, and call "k^{th} elementary transformation", the one corresponding to

$$(3.6) \qquad U = U_k \; : \; [U_k]_{ij} = u_k \, \delta_{ki} \, \delta_{ij}$$

and the label k will generally be dropped, (unless it is really necessary). The off-diagonal terms in (3.5b) yield the conditions

$$(3.7) \qquad S_{ij} \, (\lambda_j - \lambda_i) + u_k \, (\delta_{ki} \, Q_{ij} - \tilde{Q}_{ij} \, \delta_{jk}) = 0$$

so that, for $i \neq j$:

$$(3.8a) \qquad (i \neq k, \; j \neq k, \; i \neq j) \qquad S_{ij} = 0$$

$$(3.8b) \qquad (i = k, \; j \neq k) \qquad S_{kj} = u_k \, (\lambda_k - \lambda_j)^{-1} \, Q_{kj}$$

$$(3.8c) \qquad (i \neq k, \; j = k) \qquad S_{ik} = u_k \, (\lambda_k - \lambda_i)^{-1} \, \tilde{Q}_{ik}$$

Identification of the diagonal terms in (3.5c) shows that S_{ii} is constant for $i \neq k$, and that S_{kk} can be calculated by a simple integration :

$$(3.9a) \qquad (i \neq k) \quad S_{ii} = s_i$$

$$(3.9b) \qquad \frac{d}{dx} \, S_{kk}(x) = u_k \, \sum_{\ell \neq k} \frac{\tilde{Q}_{k\ell}(x) \, \tilde{Q}_{\ell k}(x) - Q_{k\ell}(x) \, Q_{\ell k}(x)}{\lambda_k - \lambda_\ell}$$

The formula (3.9b) is not convenient for the following calculations. One has better noticing that it follows from (3.1) that

$$(3.10) \qquad \frac{\partial}{\partial x} \, \det \Psi = \xi \, (\sum_n \lambda_n) \, \det \Psi$$

Since $\overset{\sim}{\Psi}$ is itself a solution of (3.1), it must also satisfy (3.10). Because of (3.2), det T is therefore constant. T has a particular form (first diagonal plus cross fixed at S_{kk}) and thus det T can be calculated by induction. The result is

$$(3.11) \qquad \det T = S_{kk}(x) + u_k \xi - u_k^2 \sum_{p \neq k} \frac{Q_{kp} \tilde{Q}_{pk}}{s_p (\lambda_k - \lambda_p)^2}$$

If we set $\det T - u_k \xi = u_k s_k$, we obtain

$$(3.12) \qquad S_{kk}(x) = u_k s_k + u_k^2 \sum_{p \neq k} s_p^{-1} (\lambda_k - \lambda_p)^{-2} Q_{kp} \tilde{Q}_{pk}$$

We are now ready to calculate the off-diagonal terms in (3.5c) :

$$(3.13) \qquad \frac{\partial}{\partial x} S_{ij} + [S_{ii} Q_{ij} - \tilde{Q}_{ij} S_{jj}] + \sum_{\ell \neq i,j} [S_{i\ell} Q_{\ell j} - \tilde{Q}_{i\ell} S_{\ell j}] = 0$$

From (3.13), using (3.8, 3.9), we obtain for $i \neq k$, $j \neq k$

$$(3.14) \qquad s_j \tilde{Q}_{ij} = s_i Q_{ij} + u_k \tilde{Q}_{ik} Q_{kj} \frac{\lambda_i - \lambda_j}{(\lambda_k - \lambda_i)(\lambda_k - \lambda_j)}$$

which enables us to get readily $\tilde{Q}_{ij}$ $(j \neq k)$ from $\tilde{Q}_{ik}$. Using now (3.8, 3.9, 3.12) in (3.13), we obtain for $j \neq k$:

$$(3.15) \qquad s_j \tilde{Q}_{kj} = \frac{u_k}{\lambda_k - \lambda_j} \frac{\partial}{\partial x} Q_{kj} + u_k \sum_{\ell \neq k} \frac{Q_{k\ell} Q_{\ell j}}{\lambda_k - \lambda_\ell}$$

$$+ u_k Q_{kj} [s_k + u_k \sum_{p \neq k} \frac{Q_{kp} \tilde{Q}_{pk}}{s_p (\lambda_k - \lambda_p)^2}]$$

which enables us to get readily $\tilde{Q}_{kj}$ $(j \neq k)$ from a knowledge of all $\tilde{Q}_{jk}$ $(j \neq k, i = 1, 2, \ldots, n)$.

Hence, it is sufficient to derive the parameters $\tilde{Q}_{jk}$ $(j \neq k)$. They appear in the last off-diagonal terms of (3.5c), which yield :

$$(3.16) \qquad \frac{u_k}{\lambda_k - \lambda_j} \frac{\partial}{\partial x} \tilde{Q}_{jk} - u_k [\tilde{Q}_{jk} s_k(x) + \sum_{\ell \neq j,k} \frac{\tilde{Q}_{j\ell} \tilde{Q}_{\ell k}}{\lambda_k - \lambda_\ell}] + s_j Q_{jk} = 0$$

where

$$(3.17) \qquad s_k(x) = s_k + u_k \sum_{p \neq k} \frac{Q_{kp} \tilde{Q}_{pk}}{s_p (\lambda_k - \lambda_p)^2}$$

The formulas (3.16) are a system of coupled Riccati equations. It is more conveniently written by using (3.14) to calculate $\tilde{Q}_{j\ell}$. The result is

$$(3.18) \qquad \frac{u_k}{\lambda_k - \lambda_j} \frac{\partial}{\partial x} \tilde{Q}_{jk} - u_k \left[s_k + u_k \sum_{p \neq k} \frac{Q_{kp} \tilde{Q}_{pk}}{s_p (\lambda_k - \lambda_p)(\lambda_k - \lambda_j)} \right] \tilde{Q}_{jk}$$

$$- u_k \sum_{\ell \neq k} \frac{s_j \, Q_{j\ell} \, \tilde{Q}_{\ell k}}{s_\ell \, (\lambda_k - \lambda_\ell)} + s_j \, Q_{jk} = 0$$

Setting then the following representation of $\tilde{Q}_{jk}$:

$$(3.19) \qquad \tilde{Q}_{jk} = -s_j \, (\lambda_k - \lambda_j) \, \varphi_j(x) \, [u_k \, \varphi_k(x)]^{-1}$$

we see that the equation (3.18) reduces to

$$(3.20) \qquad \varphi_j \left[\frac{\partial \varphi_k}{\partial x} + s_k \lambda_k \varphi_k - \sum_p Q_{jp} \varphi_p \right] - \varphi_k \left[\frac{\partial \varphi_j}{\partial x} + s_k \lambda_j \varphi_j - \sum_\ell Q_{k\ell} \varphi_\ell \right] = 0$$

Hence the system of Riccati equations (3.18) is "linearized" into the generalized Zacharov-Shabat system :

$$(3.21) \qquad \frac{\partial \varphi}{\partial x} + s_k \, J \, \varphi - Q \, \varphi = 0$$

It is important to realize that all these constructions, and the transformations themselves, proceed through "local" operators in x. Starting from a bounded norm matrix Q, elementary transformations enable us to construct a matrix $\tilde{Q}$ for which the system (3.1) is exactly solvable whereas $\tilde{Q}$ is not necessarily bounded. This can easily be checked in the simplest Zacharov-Shabat case.

Combination of elementary transformations

From (3.2) and (3.3), we check that if T_1 is a transformation which satisfies (3.3) for Q, $Q_1 = \tilde{Q}$, and if T_2 is a transformation which satisfies (3.3) for Q_1, $Q_2 = \tilde{Q}_1$, then $T_2 T_1$ is a transformation which satisfies (3.3) for Q, Q_2. Combining a trivial transformation (3.4) with an elementary transformation enables one to suppress u_k in the formulas — a cheap reduction. It is easy to combine two ξ-linear elementary transformation — say $T_\ell T_k$. If $k \neq \ell$, one can check that $T_\ell T_k$ is a ξ-linear transformation whose matrix U has only two non vanishing elements, $u_\ell s_\ell$ and $u_k \tilde{s}_k$. A ξ-linear transformation with arbitrary elements in the diagonal matrix U can be constructed by means of a suitable combination of elementary transformation. Other combinations yield ξ-polynomial transformations. Hence the elementary transformations are really the bricks any transformation of the kind we study is made of.

Transformations in the "space E"

It is known that a scattering matrix can be defined only in particular cases, e.g. the Zacharov-Shabat problem (Equation (3.1) with $n = 2$) or Schrödinger problems, and only if Q is well-behaved (finite $||Q||$). When a scattering matrix $\underset{\sim}{S}$ does exist, it is generally defined from two linearly independent matrix-solution of (3.1), say F^+ and F^-, defined themselves by their values (resp. behaviors) at points x^+ and x^- (resp. $+\infty$ and $-\infty$). The definition formula being $F^- = F^+ S$, if T^+ and T^- are the values of a transformation T at points x^+ and x^-, one readily gets from (3.2) :

$$(3.22) \qquad \tilde{S} = T^+ \, S(T^-)^{-1}$$

Hence, in the case of elementary ξ-linear transformations, if Q goes to zero at x^+ and x^-, the effect of T on the elements of S is a simple multiplication or division by a 1st degree ξ-polynomial. But this does not mean that $\tilde{Q}$ is well-behaved, and, in fact, well-known theorems [10] prevent it to be the case in the Zacharov-Shabat problem. A good behavior may be restored by a suitable combination of two linear elementary transformations, their couple being then considered as a set. The Darboux transformations cited in § 2 give an example of such a structure (a one parameter composite factor $(k-\xi) (k+\xi)^{-1}$ is introduced by the transformation), which corresponds in addition to special matrices Q and where transformation formulas are explicit.

I will not consider these cases, my aim in the present lecture being rather to emphasize the fact that transformations usually go out of the usual working classes and therefore yield a way to generalize inverse methods or to look for ambiguities, singularities, etc, in Inverse Problems.

For finite $||Q||$, Beals and Coifman have shown [9] the existence of a spectral transform of Q, which can be taken as an element of E. It is defined from the derivative of a canonical solution of (3.1) with respect to ξ (ie from its defects of holomorphy). It is easy to see that the effect of an elementary transformation is simply to add (or to remove, or to modify the coefficient of) a δ-function in this spectral transform.

Acknowledgements.   Much of this work has been done in collaboration with A. Degasperis. I also wish to thank Mrs J. Cellier for her untimely typing of the manuscript.

REFERENCES

[1] P.C. SABATIER, Well-posed Questions and Exploration of the Space of
     Parameters in linear and non-linear Inversion. pp. 82-103 in "Inverse
     Problems of Acoustic and Elastic Waves" F. Santosa et al. Ed. SIAM
     Philadelphia 1984.

[2] P.C. SABATIER, Rational reflection coefficients in One-dimensional
     Inverse Scattering and Applications. pp. 75-99 in "Conference on Inverse
     Scattering : Theory and Application" J. Bee Bednar et al. Ed. SIAM
     Philadelphia 1983.

[3] A. DEGASPERIS and P.C. SABATIER, Extension of the One-dimensional Scat-
     tering Theory, and Ambiguities. Preprint Laboratoire Physique Mathémati-
     que Montpellier PM 86-05 to be published in "Inverse Problems".

[4] J. KAY and H.E. MOSES, Inverse Scattering Papers : 1955-1963 MATH SCI
     PRESS 1982.

[5] L.D. FADDEEV, Properties of the S-matrix of the one-dimensional Schrö-
     dinger Equation. Trudy Mat. Inst. Steklov $\underline{73}$, 314-336 (1964) trans. in
     Amer. Math. Soc. Transl. (2) $\underline{65}$.

[6] P. DEIFT and E. TRUBOWITZ, Inverse Scattering on the Line Comm. Pure
     Appl. Math. $\underline{32}$, 121-251 (1979).

[7] P.B. ABRAHAM, B. DE FACIO and H.E. MOSES, Two distinct local Potentials
     with no Bound States can have the same Scattering Operator : a non uni-
     queness in Inverse Spectral Transformations Phys. Rev. Lett. $\underline{46}$,
     1657-1659 (1981).

[8] K.R. BROWNSTEIN, Non Uniqueness of the Inverse Scattering Problem and
     the Presence of E = O Bound States Phys. Rev. D 25 2704-2705.

[9] R. BEALS and R.R. COIFMAN, Scattering and Inverse Scattering for first
     order Systems Comm. Pure Appl. Math. $\underline{37}$, 39-90 (1984).

     R. BEALS and R.R. COIFMAN, Scattering, Transformations Spectrales et
     Equations d'Evolution non linéaires. Séminaire Goulaouic-Meyer-Schwartz
     1980-1981 Exposé n° 22.

[10] F. CALOGERO and A. DEGASPERIS, Spectral Transforms and Solitons tom.II,
     North Holland, in preparation.

FSC
www.fsc.org
MIX
Papier aus verantwortungsvollen Quellen
Paper from responsible sources
FSC® C105338